RSME Springer Series

Volume 8

As of 2015, RSME - Real Sociedad Matemática Española - and Springer cooperate in order to publish works by authors and volume editors under the auspices of a co-branded series of publications including advanced textbooks, Lecture Notes, collections of surveys resulting from international workshops and Summer Schools, SpringerBriefs, monographs as well as contributed volumes and conference proceedings. The works in the series are written in English only, aiming to offer high level research results in the fields of pure and applied mathematics to a global readership of students, researchers, professionals, and policymakers.

Maria Aguareles · Francesc Font · Tim Myers ·
Marta Pellicer · Joan Solà-Morales
Editors

Applications of Industrial Mathematics

158th European Study Group with Industry, Barcelona, Spain, January 27–31, 2020

Real Sociedad Matemática
Española

Springer

Editors
Maria Aguareles
Departamento de Informática, Matemática Aplicada y Estadística
Universitat de Girona
Girona, Spain

Tim Myers
Centre de Recerca Matemática
Barcelona, Spain

Joan Solà-Morales
Departament de Matemàtiques
Universitat Politècnica de Catalunya
Barcelona, Spain

Francesc Font
Universitat Politècnica de Catalunya
Barcelona, Spain

Marta Pellicer
Departamento de Informática, Matemática Aplicada y Estadística
Universitat de Girona
Girona, Spain

ISSN 2509-8888 ISSN 2509-8896 (electronic)
RSME Springer Series
ISBN 978-3-031-32129-0 ISBN 978-3-031-32130-6 (eBook)
https://doi.org/10.1007/978-3-031-32130-6

This Springer imprint is published by the registered company Springer Nature Switzerland AG
The registered company address is: Gewerbestrasse 11, 6330 Cham, Switzerland

Preface

European Study Groups with Industry (ESGI) are an activity of the European Consortium for Mathematics in Industry and are an internationally recognised method of technology transfer between academic mathematicians and industry. Typically, they are 5-day workshops where academics work alongside industrial scientists on problems of direct industrial relevance. The original study groups were held in Oxford in 1968, from there they have spread around the world, with an average of six meetings being held in Europe alone every year.

The 158th European Study Group with Industry took place at the Centre de Recerca Matemàtica in Barcelona between 27 and 31 January 2020. It was the third ESGI to be held at the CRM and was one of the final presential meetings held at the centre in that year, before COVID forced new forms of interaction and collaboration.

The problems presented at the meeting were chosen to provide a wide variety of subject areas and to appeal to local academics. Specifically, the four problems tackled were:

- Estimating the difficulty level of mobile games, presented by *King*.
- Modelling the stability of human towers, presented by *Coordinadora de Colles Castelleres de Catalunya (CCCC)*.
- Fibre coating in the manufacture of clutch components, presented by *S. A. Frenos Sauleda*.
- Safe trajectories of robot arms, presented by *FEE*.

Approximately 46 people attended the meeting, the vast majority from Spain but participants also came from Algeria, Bulgaria, North Macedonia, Portugal and the UK.

The organisers would like to thank all participants for their efforts during the meeting and also to acknowledge funding provided by the Institut d'Estudis Catalans, Secció de Ciències i Tecnologia, the Càtedra Lluís A. Santaló of the University of Girona as well as all of the companies mentioned above.

Barcelona, Spain
February 2021

Tim Myers

Contents

Chapter 1
Human Towers or *Castells* Modelling

José Antunes, Ferran Brosa Planella, Arnau Dòria-Cerezo, Aleix March San José, Marta Pellicer, Antonio Rodríguez-Ferran, and Jordi Saludes

Statement of the Problem

A *castell* or human tower is a Catalan tradition unique in the world. Dating from the 19th century, it has increased in popularity in recent years. In 2010, Unesco included the castells in the list of Intangible Cultural Heritage of Humanity (see [12]). As their motto claims, castells are a combination of strength, balance, courage and *seny* (a Catalan word for good sense and calmness).

In Catalan, the word *castell* means castle, and the complexity of each of these human structures depends both on the number of people (*castellers*) forming each floor and the number of floors. Roughly speaking, and for simple structures, the higher or thinner the more difficult the *castell* is.

J. Antunes
Centro de Ciências e Tecnologias Nucleares (C2TN), Instituto Superior Técnico (IST), Universidade de Lisboa, Lisboa, Portugal

F. Brosa Planella
Mathematics Institute, University of Warwick, Coventry, UK

A. Dòria-Cerezo
Departament d'Enginyeria Elèctrica, Institut d'Organització i Control de Sistemes Industrials, Universitat Politècnica de Catalunya, Barcelona, Spain

A. March San José
Centre de Recerca Matemàtica, Bellaterra, Spain

M. Pellicer (✉)
Department d'Informàtica, Matemàtica Aplicada i Estadística, Universitat de Girona, Girona, Spain
e-mail: marta.pellicer@udg.edu

A. Rodríguez-Ferran
Laboratori de Càlcul Numèric (LaCàN), Universitat Politècnica de Catalunya, Barcelona, Spain

J. Saludes
Research Center for Supervision, Safety and Automatic Control (CS2AC), Universitat Politècnica de Catalunya, Barcelona, Spain

M. Aguareles et al. (eds.), *Applications of Industrial Mathematics*, RSME Springer Series 8, https://doi.org/10.1007/978-3-031-32130-6_1

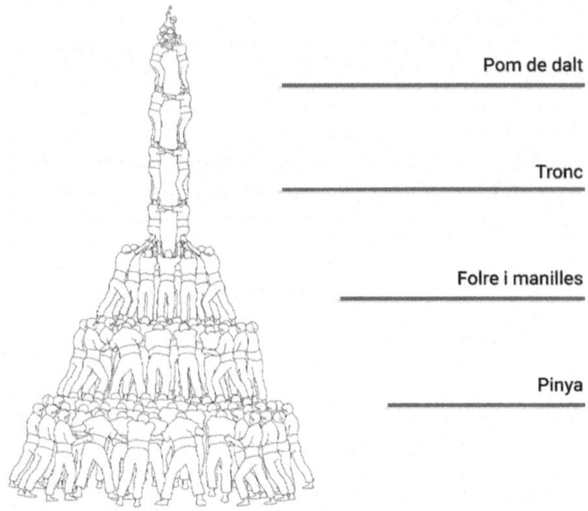

Fig. 1.1 Structure of a *castell*, indicating some of its parts. Image from https://castellscat.cat/ca/els-castells, reproduced with permission of the Coordinadora de Colles Castelleres de Catalunya

The basic structure of a *castell* consists of three parts: the *pinya* (or basis), the *tronc* (the vertical part of the castell) and the *pom de dalt* (the top three levels of the *castell*, with the *enxaneta* at the highest one). An example of the *4d8* structure is shown in Fig. 1.1 (see the Appendix A). A *castell* is accomplished if the child called *enxaneta* has climbed to the top of the *castell*, she or he has raised their hand and the *castell* has been successfully dismantled (or *descarregat*) afterwards. If the *castell* falls down once the *enxaneta* has raised his or her hand, it is said to be only *carregat*.

The *castells* are constructed in festivals throughout Catalonia, especially in the regions of Tarragona and Penedès, from where they originate. They are performed by teams or *colles*, each of them of a different town or village (or neighbourhood), that can be distinguished by the colour of their shirt. People of all the ages, genders and backgrounds form the more of the 90 *colles* that can be found in Catalonia nowadays.

The Coordinadora de Colles Castelleres de Catalunya (CCCC) is the *colles* coordinating body. One of its tasks, among others, is helping the *colles* in improving safety and prevention when constructing *castells*. One of their projects was, for instance, implementing the use of helmets in the children of the *pom de dalt*, to protect them in case of a fall.

One of the Coordinadora's current projects involves the safety of the members of the *pinya* or base when a *castell* collapses. In order to increase their protection, the Coordinadora is studying the possibility of implementing the use of a cervical protector for those *castellers* that are in the parts of the *pinya* where the *castell* is more likely to fall onto. To this purpose, they would like to have, for each type of *castell*, a "map" of the zones of the *pinya* where the *castellers* are more likely to fall onto and the intensity of this fall, as the location of the falls is not uniform. This map would also allow them to know the minimum size of the *pinya* needed for building each type of *castell* in a safer way.

From a modelling point of view, this project can be divided into two problems. The first one would consist of developing some kind of model for each type of *castell* in order to explain and study which are the mechanics that allow a *castell* to be built and, of course, which are the factors that cause a *castell* to collapse. Apart from helping in understanding the mechanics of a *castell* and the reasons of its fall, this model would also provide initial data of the second problem to be studied, which would consist of understanding how and where the *castellers* of the *tronc* and the *pom de dalt* fall down once the structure of the *castell* has broken. This second problem would provide the map of the location of the falls onto the *pinya* for each type of *castell*, which is the main objective of this project.

For more information on *castells*, see the web page of Coordinadora de Colles Castelleres de Catalunya [13], Wikipedia [14] or [6, 10] (in Catalan).

1.1 Introduction

Castells[1] are human towers (see [12–14] and Fig. 1.2) built traditionally at festivals in Catalonia, the Balearic Islands and the Valencian Community. The tradition stems from the *Ball de Valencians* in Valls (first documented in 1712) and spread over the course of the 18th century to other towns in the area. Increasing its popularity in recent years there are now around a hundred teams or *colles* around the world, most of them in Catalonia, see [13].

In a *castell* a number of people (*castellers*) build a human structure by climbing on top of each other. A *castell* succeeds when a child called *enxaneta* has climbed to the top of the *castell*, crossed the top of the structure (raising their hand at this moment) and the *castell* has been successfully disassembled (or *descarregat*) afterwards. If the *castell* falls down once the *enxaneta* has arrived at the top of it, it is said to be only *carregat*. The basic structure of a *castell* consists of three parts: the *pinya* (or base), the *tronc* (the vertical part of the *castell*) and the *pom de dalt* (the top three levels of the *castell*, with the *enxaneta* at the top), see Fig. 1.1.

Most *castells* are successful, but some of them end up collapsing. The Coordinadora de Colles Castelleres de Catalunya (CCCC), the *colles* coordinating body, has as one of its tasks to help the *colles* to improve their safety while doing *castells*. The present report is based in this area. It is part of a larger project suggested by the CCCC consisting of determining the area of the *pinya* where it is more likely that *castellers* impact when a *castell* collapses. As stated, this is important as it would allow the CCCC to increase the safety of the *pinya* members: it would allow to quantify the minimum required size of a *pinya* for preventing people falling directly on the floor, and also to determine the most *dangerous* areas to prescribe, if necessary, additional external protection devices (such as spine protection elements, currently under study).

[1] The emphasized terms are technical terms and, hence, we have kept their form in Catalan. A Glossary with all of them is included in Appendix A.

To this end, our approach splits the above problem into the following stages (see Fig. 1.3):

- The **standing or running stage**, containing both the **stable** and **failure** stages in the building of a *castell*. It therefore includes the admissible dynamics of a *castell* when it is being built and the analysis of the limits where *castellers* are able to

Fig. 1.2 Castell performed on the Day of the Castellers of Sant Cugat in 2012. By Gausacs - Own work, CC BY-SA 4.0, https://commons.wikimedia.org/w/index.php?curid=38457139

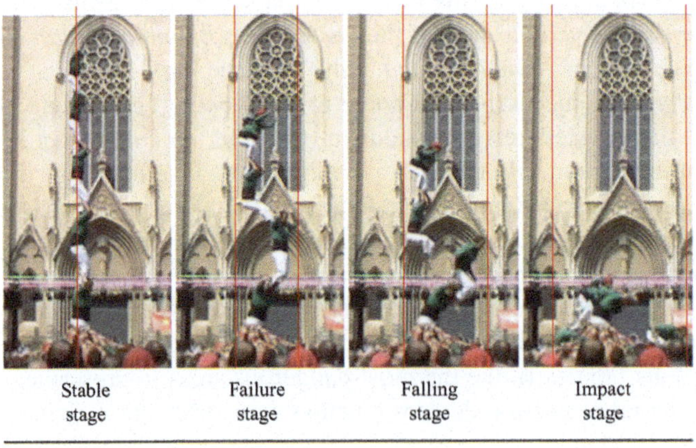

Fig. 1.3 Stages when a *castell* collapses and their associated risky areas (in red). The photo corresponds to a *pilar de 6 carregat (pd6c)*, from Castellers de Sabadell, 2012. Reproduced with permission from Gemma Martí

remain together and the mechanisms that allow it or the factors that make a *castell* to collapse.
- The **falling stage** starts when lateral forces vanish and *castellers* are only affected by gravity. Then, gravity and both plastic and elastic impacts between *castellers* need to be considered.
- Finally, the **impact stage** contains the first impact to the *pinya* and all the subsequent impacts until all *castellers* have stopped falling.

This approach allows the definition of a risky area related to each stage as well as to understand the mechanisms that are present in each, and provides the initial conditions for the next stage.

The present work is focused on the study of the running stage and the operating capability of a *castell* within it. The importance of this work is twofold: apart from understanding the dynamics that allow a *castell* to be built, we will try to capture and understand the main mechanisms and factors that allow a *castell* to be successfully constructed or to collapse. The next stages would analyse how and where the *castellers* of the *tronc* fall down, once the structure of the *castell* is broken. Although Biomechanics is a wide field of study (with plenty of references in the literature) we have to say that, to our knowledge, this is the first work on the modelling of *castells* from a mechanical point of view. In this sense, the few accurate scientific studies in the literature are provided in [6, 7].

As we explain below, in the present work the analysis is focused on the *pilar* structure, composed by just one *casteller* at each level of the *tronc*. Of course, in future works this analysis should be extended to more complex 3D structures.

The pillar: a paradigm of most modelling issues. *Castells* can display many configurations, ranging from simple *pilars*, see Fig. 1.4(A), to very complex 3D structures. After carefully scrutinizing the addressed problem, we decided to focus this preliminary work on a generic *pilar*, with N levels. Actually, the geometry of such structures is the simplest one among the *castellers* repertoire (we can think of them as 1D structures) and, moreover, they typically display near-planar motions (hence 2D movement). In spite of this simplicity, the essential features of all *castells* are present when modelling *pilars*, namely the need for (a) both static and dynamical structural models, (b) neuro-muscular control laws, (c) balancing criteria, (d) *castell* disruption criteria, and (e) a realistic post-failure dynamical formulation. Only after all these aspects have been mastered and calibrated on the simplest *pilar* structures, can one attempt to apply the developed computational strategies to more complex *castells* displaying 3D motions.

This work first starts with the dynamical approach (Sects. 1.2 and 1.3) and then focuses on a static balance approach (Sect. 1.4). This may seem contrary to the construction of the models, but we will see how this makes sense and, moreover, how it allows us to start with a 1-link *casteller* model and continue with a more realistic 3-link one.

In Sect. 1.2 we present a first dynamical model where we think of the pillar as an inverted pendulum with several links or bars (each of these links representing a *casteller*) stabilized by restoring torques that are applied to their joints. First, for

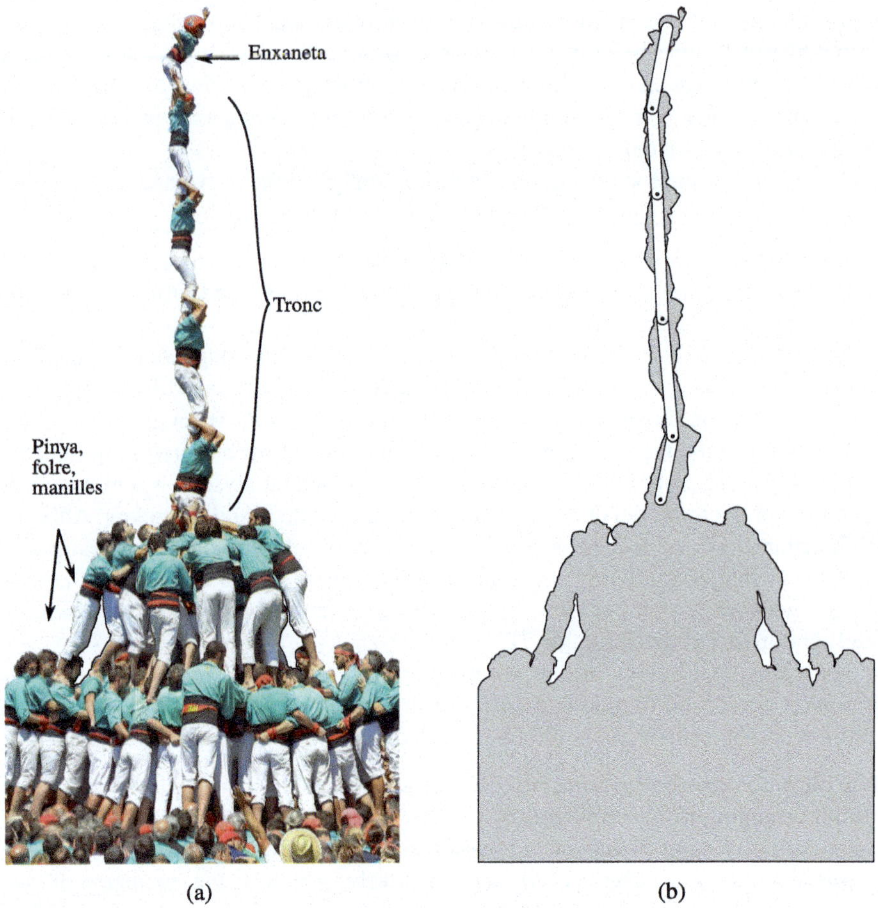

Fig. 1.4 Example of a real pilar (a) and the diagram with the corresponding approach as an N-link inverted pendulum (b). *Pilar de 8 amb folre i manilles (pd8fm)*, Castellers de Vilafranca, Mataró, 2014. (Photo: Juanma Ramos, https://www.elpuntavui.cat/societat/article/5-societat/762686-santes-extraordinaries.html. Reproduced with permission from Hermes Comunicacions SA)

the uncontrolled dynamics, the Lagrangian formulation for the problem is proposed based on the link angles and angular velocities. Although this formulation may not be new, it has to be emphasized that the point of view adopted in this section allows us to write the corresponding dynamical system in a very compact and straightforward way, which is important as the number of equations may be large.

In Sect. 1.3 we look for a more realistic model of a *casteller*, which can capture the bent-knee position often used when performing human towers. Therefore, we move from one to three links per *casteller*. This new model raises a few questions that need to be considered. The first one is the biomechanical restriction in the internal joints, as they cannot be bent beyond a certain angle. We also consider the restoring torques that allow the *casteller* to stand in its desired position as a control problem, where the reaction times of the *casteller* may be included as time delays. This leads us to consider the balance criteria imposed by a *casteller* to maintain their standing position. Finally, we study the dynamics of this bent-knee position: we run simulations for the behaviour under an initial perturbation of different positions. Some preliminary simulations seem to validate the benefits of the bent-knee position, as it allows a faster return to the desired positions, giving as well a better reaction capability to the *casteller*.

In Sect. 1.4 we take another approach. While in the previous sections we have considered a dynamical point of view (first for a 1-link *casteller* and then for a 3-link one) we now consider a static and fixed configuration for a *pilar* and compute the conditions for the restoring torques to ensure its static balance. These computations suggest some quality indices showing how far is a given configuration from collapsing (and which will be the effort to avoid it). These indices are the mass center eccentricity (from the perfectly vertical position), and what we call the folding index and the suffering index, which measure how *bent* are the joints of the *castellers* and how much effort their joint muscles are applying, respectively. As examples of each, we numerically evaluate these indicators on configurations taken from two videos of pillars (one successful, the other not).

As future considerations (among others) it is clear that accurate biomechanical data is needed to validate the approaches above and go further, as much as good positional data on real *castells*. The control problem that follows from an active *casteller* is another question that with no doubt is present in reality and, although it is considered in the present work, it should definitely be developed in a future line of research in this project. Once the current approaches are fully developed for the *pilar* structure, the next step is of course to extend and validate the model to more complex constructions. These and other conclusions, and future lines of research are considered in Sect. 1.5.

Finally, we include a glossary of the most common terms in the world of *castells*, for a better understanding of the terminology (see Appendix A).

1.2 A Basic Dynamical Point of View: The N-Link Inverted Pendulum

In this section we give a non-linear formulation of the assembled N-link pillar. As illustrated in Fig. 1.4, the *castell* configuration addressed here is the simplest possible, and mimics a *pilar* structure with an arbitrary number N of "stacked" *castellers*. Each element $i = 1, 2, \ldots, N$ is modelled through a rigid bar with total length l_i and distributed mass, with an equivalent total mass m_i "assigned" to the center of mass at local axial coordinate $G_i = c_i l_i$. If one assumes that the bars masses are uniformly distributed, then the relative local axial coordinate of the center of mass

Fig. 1.5 Scheme with the proposed notation. We recall that Γ_i, θ_i, J_i stand for the restoring moment, the rotation angle and the moment of inertia of bar i, respectively. We also recall that c_i, l_i stand for the relative location of the center of mass and the total length of the bar i, respectively

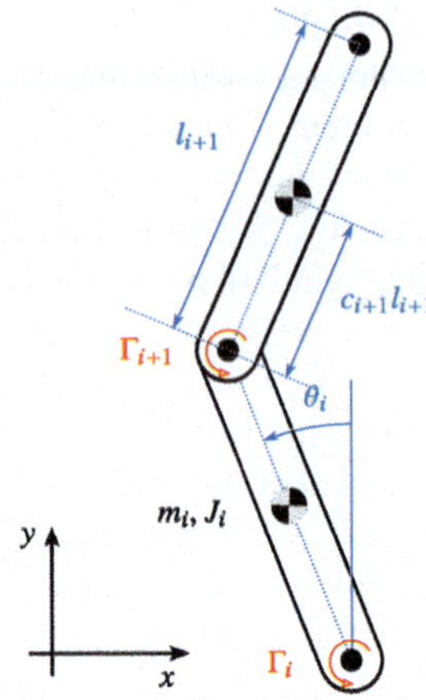

of each element is given by $c_i = G_i/l_i$, while the moment of inertia about the center of mass is $J_i = m_i l_i^2/12$.

For the fully assembled structure, the bars are articulated at their extremities, each bar being described in terms of the corresponding rotation angle $\theta_i(t)$ (measured counterclockwise from the vertical). Because any inverted pendulum is inherently unstable, in order to stabilize this N-link pendulum, it will be assumed that equivalent restoring moments $\Gamma_i(t)$ are applied by the *castellers* legs and arms at the successive articulations. The notation is depicted in Fig. 1.5.

Referring to Fig. 1.4b, the fully assembled N-link *pilar* may be formulated in terms of the generalized coordinates $\theta_i(t)$, with $i = 1, 2, \ldots, N$. The nonlinear motion equations of the inverted multiple pendulum will be developed using the Lagrange formulation, with respect to the generalized displacements $\theta_i(t)$ and velocities $\dot{\theta}_i(t)$ (see footnote[2]):

$$\frac{d}{dt}\left(\frac{\partial T}{\partial \dot{\theta}_i}\right) - \frac{\partial T}{\partial \theta_i} + \frac{\partial V}{\partial \theta_i} = \Gamma_i, \quad i = 1, 2, \ldots, N. \tag{1.1}$$

The potential energy V and the kinetic energy T are given by

$$V(t) = g\sum_{i=1}^{N} m_i y_i, \tag{1.2a}$$

$$T(t) = \frac{1}{2}\sum_{i=1}^{N}\left(m_i|v_i|^2 + J_i\dot{\theta}_i^2\right), \tag{1.2b}$$

where the locations of the centers of mass of the links are given by

[2] For the sake of simplicity, from now on and when clear from the context, the time argument (t) in the variables is omitted.

$$x_i(t) = -c_i l_i \sin\theta_i - \sum_{j=1}^{i-1} l_j \sin\theta_j, \tag{1.3a}$$

$$y_i(t) = c_i l_i \cos\theta_i + \sum_{j=1}^{i-1} l_j \cos\theta_j, \tag{1.3b}$$

and the velocities by

$$\boldsymbol{v}_i(t) = \begin{pmatrix} \dot{x}_i \\ \dot{y}_i \end{pmatrix} = -\begin{pmatrix} c_i l_i \dot{\theta}_i \cos\theta_n + \sum_{j=1}^{i-1} l_j \dot{\theta}_j \cos\theta_j \\ c_i l_i \dot{\theta}_i \sin\theta_i + \sum_{j=1}^{i-1} l_j \dot{\theta}_j \sin\theta_j \end{pmatrix}. \tag{1.4}$$

Then, we obtain from (1.2)–(1.4)

$$V(t) = g \sum_{i=1}^{N} m_i \left(c_i l_i \cos\theta_i + \sum_{j=1}^{i-1} l_j \cos\theta_j \right), \tag{1.5a}$$

$$\begin{aligned} T(t) = {} & \frac{1}{2} \sum_{i=1}^{N} m_i \left(c_i l_i \dot{\theta}_i \cos\theta_i + \sum_{j=1}^{i-1} l_j \dot{\theta}_j \cos\theta_j \right)^2 \\ & + \frac{1}{2} \sum_{i=1}^{N} m_i \left(c_i l_i \dot{\theta}_i \sin\theta_i + \sum_{j=1}^{i-1} l_j \dot{\theta}_j \sin\theta_j \right)^2 + \frac{1}{2} \sum_{i=1}^{N} J_i \dot{\theta}_i^2, \end{aligned} \tag{1.5b}$$

and, replacing (1.5) into (1.1) we can obtain the motion equations.

A convenient systematic manner to obtain the motion equations of the assembled system is provided in [2]. First, in order to state the kinematic variables, the coordinates of the centers of mass of the links are expressed in terms of the angles $\theta_i(t)$ at the joints. Then, we can write the positions as

$$\boldsymbol{x}(t) = -\boldsymbol{\Psi} \sin\boldsymbol{\theta},$$
$$\boldsymbol{y}(t) = \boldsymbol{\Psi} \cos\boldsymbol{\theta},$$

where $\boldsymbol{x}(t) = (x_1, \ldots, x_N)^T$, $\boldsymbol{y}(t) = (y_1, \ldots, y_N)^T$, $\boldsymbol{\theta}(t) = (\theta_1, \ldots, \theta_N)^T$, and with sin and cos applied element-wise to $\boldsymbol{\theta}$. The transformation matrix $\boldsymbol{\Psi}$ is defined as

$$\boldsymbol{\Psi} = \begin{pmatrix} c_1 l_1 & 0 & 0 & \cdots & 0 & 0 \\ l_1 & c_2 l_2 & 0 & \cdots & 0 & 0 \\ l_1 & l_2 & c_3 l_3 & \cdots & 0 & 0 \\ \vdots & \vdots & \vdots & \ddots & \vdots & \vdots \\ l_1 & l_2 & l_3 & \cdots & c_{N-1} l_{N-1} & 0 \\ l_1 & l_2 & l_3 & \cdots & l_{N-1} & c_N l_N \end{pmatrix}.$$

Then, the velocities of the links' centers of mass are computed from the time-derivatives

$$\dot{\boldsymbol{x}} = -\boldsymbol{\Psi}\frac{\mathrm{d}}{\mathrm{d}t}(\sin\boldsymbol{\theta}) = -\boldsymbol{\Psi}\mathbf{D}_{\mathrm{c}}\dot{\boldsymbol{\theta}},$$
$$\dot{\boldsymbol{y}} = \boldsymbol{\Psi}\frac{\mathrm{d}}{\mathrm{d}t}(\cos\boldsymbol{\theta}) = -\boldsymbol{\Psi}\mathbf{D}_{\mathrm{s}}\dot{\boldsymbol{\theta}},$$

with the $N \times N$ diagonal matrices $\mathbf{D}_{\mathrm{c}}(t) = \mathrm{diag}(\cos\boldsymbol{\theta})$ and $\mathbf{D}_{\mathrm{s}}(t) = \mathrm{diag}(\sin\boldsymbol{\theta})$.

Now, from the previous results, we can compute the potential and kinetic energies in matrix/vector form

$$V(t) = g\boldsymbol{m}^T\boldsymbol{y},$$
$$T(t) = \frac{1}{2}\left(\dot{\boldsymbol{x}}^T\mathbf{D}_m\dot{\boldsymbol{x}} + \dot{\boldsymbol{y}}^T\mathbf{D}_m\dot{\boldsymbol{y}} + \dot{\boldsymbol{\theta}}^T\mathbf{D}_J\dot{\boldsymbol{\theta}}\right),$$

hence

$$V(t) = g\boldsymbol{m}^T\boldsymbol{\Psi}\cos\boldsymbol{\theta}, \tag{1.9a}$$
$$T(t) = \frac{1}{2}\dot{\boldsymbol{\theta}}^T\left(\mathbf{D}_{\mathrm{c}}\boldsymbol{\Psi}^T\mathbf{D}_m\boldsymbol{\Psi}\mathbf{D}_{\mathrm{c}} + \mathbf{D}_{\mathrm{s}}\boldsymbol{\Psi}^T\mathbf{D}_m\boldsymbol{\Psi}\mathbf{D}_{\mathrm{s}} + \mathbf{D}_J\right)\dot{\boldsymbol{\theta}}, \tag{1.9b}$$

with $\boldsymbol{m} = (m_1, \ldots, m_N)^T$, $\mathbf{D}_m = \mathrm{diag}(\boldsymbol{m})$ and $\mathbf{D}_J = \mathrm{diag}(J_1, \ldots, J_N)$.

Then, the dynamical equations are obtained replacing (1.9) in to the Lagrange equation in vector form

$$\frac{\mathrm{d}}{\mathrm{d}t}\left(\frac{\partial T}{\partial\dot{\boldsymbol{\theta}}}\right) - \frac{\partial T}{\partial\boldsymbol{\theta}} + \frac{\partial V}{\partial\boldsymbol{\theta}} = \boldsymbol{\Gamma},$$

with the vector of restoring torques defined as $\boldsymbol{\Gamma}(t) = (\Gamma_1, \ldots, \Gamma_N)^T$.

For details of the matrix/vector derivative computations see [8]. The following nonlinear dynamical formulation is obtained for the assembled pillar

$$\mathbf{M}(\boldsymbol{\theta})\ddot{\boldsymbol{\theta}} + \mathbf{D}(\boldsymbol{\theta})\dot{\boldsymbol{\theta}}^2 + \boldsymbol{g}(\boldsymbol{\theta}) = \boldsymbol{\Gamma}. \tag{1.10}$$

Hence, using a more convenient notation we can write

$$\mathbf{M}(\boldsymbol{\theta})\ddot{\boldsymbol{\theta}} + \mathbf{C}(\dot{\boldsymbol{\theta}}, \boldsymbol{\theta})\dot{\boldsymbol{\theta}} + \boldsymbol{g}(\boldsymbol{\theta}) = \boldsymbol{\Gamma}, \tag{1.11}$$

where

$$\mathbf{M}(\boldsymbol{\theta}) = \mathbf{D}_{\mathrm{c}}\boldsymbol{\Psi}^T\mathbf{D}_m\boldsymbol{\Psi}\mathbf{D}_{\mathrm{c}} + \mathbf{D}_{\mathrm{s}}\boldsymbol{\Psi}^T\mathbf{D}_m\boldsymbol{\Psi}\mathbf{D}_{\mathrm{s}} + \mathbf{D}_J,$$
$$\mathbf{C}(\dot{\boldsymbol{\theta}}, \boldsymbol{\theta}) = \mathbf{D}(\boldsymbol{\theta})\mathrm{diag}(\dot{\boldsymbol{\theta}}) = \left[\mathbf{D}_{\mathrm{s}}\boldsymbol{\Psi}^T\mathbf{D}_m\boldsymbol{\Psi}\mathbf{D}_{\mathrm{c}} - \mathbf{D}_{\mathrm{c}}\boldsymbol{\Psi}^T\mathbf{D}_m\boldsymbol{\Psi}\mathbf{D}_{\mathrm{s}}\right]\mathrm{diag}(\dot{\boldsymbol{\theta}}),$$
$$\boldsymbol{g}(\boldsymbol{\theta}) = -g\mathbf{D}_{\mathrm{s}}\boldsymbol{\Psi}^T\boldsymbol{m}.$$

One can notice several points in (1.10):

1. The first term, of inertial nature, with a global mass matrix which depends on the instantaneous configuration $\boldsymbol{\theta}(t)$. $\mathbf{M}(\boldsymbol{\theta})$ is known as the inertia matrix.
2. The second term, which depends on both the instantaneous configuration $\boldsymbol{\theta}(t)$ and velocity $\dot{\boldsymbol{\theta}}(t)$, encapsulates the centrifugal and Coriolis-type terms.
3. The third term stems from the gravity potential, depending on the instantaneous configuration $\boldsymbol{\theta}(t)$, is of stiffness nature.

From (1.11) the acceleration can be computed at each time-step as

$$\ddot{\boldsymbol{\theta}} = \mathbf{M}(\boldsymbol{\theta})^{-1}\left(\boldsymbol{\Gamma} - \mathbf{C}(\dot{\boldsymbol{\theta}}, \boldsymbol{\theta})\dot{\boldsymbol{\theta}} - \boldsymbol{g}(\boldsymbol{\theta})\right),$$

enabling the numerical time-step integration with any suitable algorithm. Here, a Newmark scheme was used.

1.3 Enhancing the Dynamical Model: A 3-Link *Casteller* Model

If we look at a picture of a *pilar*, like in Fig. 1.4, we observe that the *castellers* do not stand in a vertical position but they bend their knees and hips a little bit (which we refer to as the bent-knee position). This is observed, at a higher or lower degree, in all *castells*, but it is especially noticeable in *pilars*. Therefore, this bending of the joints suggests that modelling each *casteller* as a single rigid link might not be an accurate description of reality and a *casteller* should be composed of several links. Then, one can think that the *pilar* can be directly represented by (1.11), just taking $L \cdot P$ links where L and P are the number of links (per *casteller*) and *castellers*, respectively.

In the Biomechanics literature we can find multiple ways to describe a human body as a set of articulated rigid bars [3, 9]. However, looking back at Fig. 1.4, we see that, from the structural point of view, each *casteller* can be represented by three main links: one for the lower legs, one for the upper legs and one for the body. See Fig. 1.6, where *castellers* are i-labelled and each link takes values in $j = 1, 2, 3$, representing lower leg, upper leg and body, and their corresponding joints (feet, knees, hip), respectively. Additionally, a closer look shows that there are some complicating issues that arise from considering each *casteller* as a 3-link structure and they will be discussed below.

Along this section, as only a single *casteller* is considered, we omit the sub-index i. The aim of this section is to present a simple but more realistic model, including self control, for understanding how a *casteller* behaves when performing a *castell*. This single 3-link *casteller* model is aimed to be a stepping-stone towards modelling a *castell* composed of 3-link *castellers*.

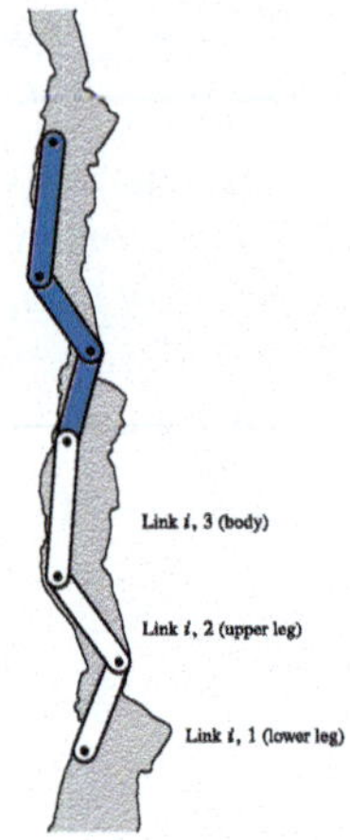

Fig. 1.6 Scheme of two *castellers* represented by the 3-link model

A Single 3-Link Casteller

First, let us obtain the model for a 3-link *casteller*. From the general dynamics obtained in (1.11), and taking $N = 3$, a single *casteller* can be represented by the nonlinear equations

$$\mathbf{M}(\boldsymbol{\theta})\,\ddot{\boldsymbol{\theta}} + \mathbf{C}(\dot{\boldsymbol{\theta}}, \boldsymbol{\theta})\dot{\boldsymbol{\theta}} + \boldsymbol{g}(\boldsymbol{\theta}) = \boldsymbol{\Gamma}_u, \tag{1.12}$$

where $\boldsymbol{\Gamma}_u$ represents now the joint moments of the *casteller* reacting to the external disturbances, and the inertia, Coriolis and potential related matrices result, respectively, in

$$\mathbf{M}(\boldsymbol{\theta}) = \begin{pmatrix} b_{11} & b_{12}\cos(\theta_1 - \theta_2) & b_{13}\cos(\theta_1 - \theta_3) \\ b_{12}\cos(\theta_1 - \theta_2) & b_{22} & b_{23}\cos(\theta_2 - \theta_3) \\ b_{13}\cos(\theta_1 - \theta_3) & b_{23}\cos(\theta_2 - \theta_3) & b_{33} \end{pmatrix},$$

$$\mathbf{C}(\dot{\boldsymbol{\theta}}, \boldsymbol{\theta}) = \begin{pmatrix} 0 & b_{12}\dot{\theta}_2\sin(\theta_1 - \theta_2) & b_{13}\dot{\theta}_3\sin(\theta_1 - \theta_3) \\ -b_{12}\dot{\theta}_1\sin(\theta_1 - \theta_2) & 0 & b_{23}\dot{\theta}_3\sin(\theta_2 - \theta_3) \\ -b_{13}\dot{\theta}_1\sin(\theta_1 - \theta_3) & -b_{23}\dot{\theta}_2\sin(\theta_2 - \theta_3) & 0 \end{pmatrix},$$

$$\boldsymbol{g}(\boldsymbol{\theta}) = -g \begin{pmatrix} l_1(c_1 m_1 + m_2 + m_3)\sin(\theta_1) \\ l_2(c_2 m_2 + m_3)\sin(\theta_2) \\ l_3 c_3 m_3 \sin(\theta_3) \end{pmatrix},$$

with coefficients

$$
\begin{aligned}
b_{11} &= l_1^2(c_1^2 m_1 + m_2 + m_3) + J_1, \\
b_{22} &= l_2^2(c_2^2 m_2 + m_3) + J_2, \\
b_{33} &= l_3^2 c_3^2 m_3 + J_3, \\
b_{12} &= l_1 l_2 (c_2 m_2 + m_3), \\
b_{13} &= l_1 l_3 c_3 m_3, \\
b_{23} &= l_2 l_3 c_3 m_3.
\end{aligned}
$$

The reactions that a *casteller* does in order to avoid the *pilar* from collapsing can be seen from a mathematical point of view as stability and control problems. There exist many control algorithms for systems with the form (1.12). Among them, passivity-based controllers ensures asymptotic convergence to the desired trajectory, and have been used in many applications such as robot manipulators [5]. Then, the reactions of the *casteller*, can be described as control actions with the form

$$
\boldsymbol{\Gamma}_u = \hat{\mathbf{M}}(\boldsymbol{\theta})(\ddot{\boldsymbol{\theta}}_d - \mathbf{K}_d(\dot{\boldsymbol{\theta}} - \dot{\boldsymbol{\theta}}_d) - \mathbf{K}_p(\boldsymbol{\theta} - \boldsymbol{\theta}_d)) + \hat{\mathbf{C}}(\dot{\boldsymbol{\theta}}, \boldsymbol{\theta})\dot{\boldsymbol{\theta}} + \hat{\mathbf{g}}(\boldsymbol{\theta}), \tag{1.13}
$$

where $\boldsymbol{\theta}_d$ (and their respective time derivatives) stand for the desired angular positions, velocities and accelerations, and matrices

$$
\mathbf{K}_d = \mathrm{diag}(K_{d1}, K_{d2}, K_{d3},), \mathbf{K}_p = \mathrm{diag}(K_{p1}, K_{p2}, K_{p3},)
$$

are the gain matrices that have to be tuned according the desired dynamic response and can be physically related to a damping and elastic effects, respectively, and matrices $\hat{\mathbf{M}}(\boldsymbol{\theta})$, $\hat{\mathbf{C}}(\dot{\boldsymbol{\theta}}, \boldsymbol{\theta})$, $\hat{\mathbf{g}}(\boldsymbol{\theta})$ are the estimated ones from the knowledge and perception that each *casteller* have from her/him self. In this case the *casteller* aims for an steady position, so that the desired velocities and accelerations are null, $\dot{\boldsymbol{\theta}}_d = \ddot{\boldsymbol{\theta}}_d = \mathbf{0}$.

The resulting (closed-loop) dynamics of the *casteller* is obtained replacing (1.13) in (1.12). In ideal conditions, when estimated matrices in (1.13) match with the actual values in (1.12), the closed-loop dynamics simplifies in the decoupled second order dynamics for each angle

$$
(\ddot{\boldsymbol{\theta}} - \ddot{\boldsymbol{\theta}}_d) + \mathbf{K}_d(\dot{\boldsymbol{\theta}} - \dot{\boldsymbol{\theta}}_d) + \mathbf{K}_p(\boldsymbol{\theta} - \boldsymbol{\theta}_d) = 0,
$$

where $\mathbf{K}_d, \mathbf{K}_p$ unequivocally define the response. Any deviation of the estimated values from the actual ones will affect the performance of the *casteller* and also its stability.

In this work, as a first approximation and using the previous considerations, the proposed controller has been simplified approximating $\theta_j \approx 0$ in $\hat{\mathbf{M}}(\boldsymbol{\theta})$ and $\hat{\mathbf{C}}(\boldsymbol{\theta})$ (that implies $\hat{\mathbf{M}}(\boldsymbol{\theta}) \approx \mathbf{I}$ and $\hat{\mathbf{C}}(\dot{\boldsymbol{\theta}}, \boldsymbol{\theta}) \approx \mathbf{0}$), and (1.13) results in

$$
\boldsymbol{\Gamma}_u = -\mathbf{K}_d\dot{\boldsymbol{\theta}} - \mathbf{K}_p(\boldsymbol{\theta} - \boldsymbol{\theta}_d) + \mathbf{g}(\boldsymbol{\theta}). \tag{1.14}
$$

Notice that the latter simplification is valid for a *casteller* in a complete vertical position. However, as has been previously discussed, we are aiming for representing each *casteller* in a bent-knee position, that means $\boldsymbol{\theta}_d \neq 0$ and, consequently, $\mathbf{M}(\boldsymbol{\theta}_d) \neq \mathbf{I}$. This suggests using (1.13) in a future.

Kinematic Constraint for the Joints

If we consider a multiple link *casteller*, then we must take into account the different types of joints involved in the problem. We need to distinguish between the internal joints of a *casteller* (namely knees and hip) and the external joint between two different *castellers*, which is the system comprised by the ankles of the top *casteller* and the arms of the bottom *casteller*. One of the key differences between the internal and external joints is that the internal ones show an asymmetric behaviour. Even though that all the joints show kinematic constraints if pushed to extreme enough cases, the internal joints are reasonably close to these constraints. For example, starting from a perfectly vertical position, the hip can only bend forward while the knees can only bend backwards. This limitation can be modelled adding an extra moment vector, $\boldsymbol{\Gamma}_{\text{kin}}$, so that (1.12), results into

$$\mathbf{M}(\boldsymbol{\theta})\,\ddot{\boldsymbol{\theta}} + \mathbf{C}(\dot{\boldsymbol{\theta}}, \boldsymbol{\theta})\dot{\boldsymbol{\theta}} + \boldsymbol{g}(\boldsymbol{\theta}) = \boldsymbol{\Gamma}_u + \boldsymbol{\Gamma}_{\text{kin}},$$

where the kinematic constraint can be written as a nonlinear restoring torque defined as

$$\Gamma_{\text{kin,j}} = \max\left(\mp K_{\text{kin,j}}(\theta_{j+1} - \theta_j - \theta_{\text{lim},j}), 0\right),$$

where θ_j and θ_{j+1} are the angles of the two links connected at the joint, K_{kin} is the stiffness of the constraint (which should be very high), θ_{lim} is the limit angle beyond which the joint cannot bend, and the sign depends on which side of the limit angle $\theta_{\text{lim},j}$ is restricted.

In future work, further differences will arise when the collapse of a *castell* is considered, as the external joints may break but the internal ones cannot.

Desired Position

Another aspect to consider when increasing the number of links in a *casteller* is the balance criterion, which is closely related to the controller. When a *casteller* is composed of a single link, it is logical to impose that the balance criterion is to aim for a vertical position. However, when we consider 3-link *castellers* the balance criterion is less obvious. In this section we discuss some of the balance criteria that are taken into account from a single *casteller* point of view in order to maintain the *pilar* to stand, but a further discussion on these mechanisms from a collective point of view of the whole *pilar* (that is, interaction among *castellers* to avoid the *castell* to collapse) will be given in Sect. 1.4. Linking both approaches is, of course, something that should be considered in a future work.

The first balance condition is that the centre of mass of a given *casteller* is vertically aligned with their feet. The physical explanation of this constraint is that a *casteller* aims to reduce the moment applied to the external joint caused by their own weight. Mathematically, this condition can be written as $d = 0$ with

$$d = -\frac{1}{\sum_{j=1}^{3} m_j} \sum_{j=1}^{3} m_j \left(c_j l_j \sin\theta_j(t) + \sum_{k=1}^{j-1} l_k \sin\theta_k(t) \right), \tag{1.15}$$

where for each bar i m_i is the mass, c_i is the position of the centre of mass, l_i the length and θ_i the rotation angle of the bar. This condition is closely related with the *mass center eccentricity* described in Sect. 1.4. Actually, formula (1.15) will be generalized in formula (1.21) when taking into account the displacement of the *castellers* below (see Sect. 1.4 for more details).

The second balance condition is that the shoulders of a given *casteller* are vertically aligned with their feet too. The physical explanation of this constraint is that a *casteller* aims to align the force received from the *castellers* above with their feet to reduce the moment generated. Mathematically, this condition can be written as $w = 0$ with

$$w = -\sum_{j=1}^{3} l_j \sin\theta_j(t). \tag{1.16}$$

Finally, the last balance condition is that the centre of mass of a given *casteller* is at a given height. Mathematically, this condition can be written as $h = h_d$ with

$$h = \frac{1}{\sum_{j=1}^{3} m_j} \sum_{j=1}^{3} m_j \left(c_j l_j \cos\theta_j(t) + \sum_{k=1}^{j-1} l_k \cos\theta_k(t) \right), \tag{1.17}$$

and h_d being a desired position. The physical interpretation of this last condition is less clear but it represents the fact that certain bent-knee positions are more comfortable to sustain for long times than others. The comfortable position is described in this case by h_d. This condition is closely related with the *folding index* described in Sect. 1.4.

Defining the position of a *casteller* in a compact way as

$$\boldsymbol{\xi}(\boldsymbol{\theta}) = \begin{pmatrix} d(\boldsymbol{\theta}) \\ w(\boldsymbol{\theta}) \\ h(\boldsymbol{\theta}) \end{pmatrix} \tag{1.18}$$

the new variable $\boldsymbol{\xi}$ provides the necessary change of coordinates to obtain the desired steady angles as

$$\boldsymbol{\theta}_d = \boldsymbol{\xi}^{-1}(0, 0, h_d). \tag{1.19}$$

A further analysis on the existence of (1.19) would be necessary to ensure a set of admissible target angles without violating physical constraints. Also, the results of Sect. 1.4 could help in this analysis.

The desired position in terms of angles (1.19) can be directly used in (1.13) and the *casteller* will reach the desired position. However, this control strategy does not coordinate the angles in order to reach the target position, which makes it not very realistic. For a more accurate model, it is necessary to provide reactions of the *casteller* able to look for the balance conditions defined as (1.15), (1.16) and (1.17). As we said, that is something that will be considered in Sect. 1.4, but not from a control point of view. Hence, an interesting future work would be to find a more realistic controller able to simulate better the coordinated reactions of the *casteller* at each joint by analysing the dynamics of $\boldsymbol{\xi}$ in (1.18).

Delay-differential Balance Model

Another effect that could be taken into account is the delay in the human-induced restoring moments (see [1, 4]). The delay represents the reaction time of a *casteller* when trying to recover the balance. In this case, we can write the general form of the controller as

$$\boldsymbol{\Gamma}(t) = f(\dot{\boldsymbol{\theta}}(t-\tau), \boldsymbol{\theta}(t-\tau)), \tag{1.20}$$

where f is the controller function, which could be set to have a similar form to (1.13) or (1.14), and τ is the delay parameter. The interpretation of (1.20) is that the reaction of the *casteller* at time t is based on the position at time $t-\tau$ where the delay parameter represents the reaction time. Note that in (1.20) we assume that the delay parameter is the same for all the joints, however it could be reformulated to account for different delay parameters and, in the future, for different *castellers* (according to their biomechanical characteristics and expertise). In addition, we could assume that the delay parameter changes in time, as *castellers* become progressively more fatigued.

Preliminary simulations of an inverted pendulum modelling a pillar (with *castellers* represented by a single link model) have shown that delays associated to neuromuscular behaviours, reaction times and fatigue, could dramatically compromise the stability of the structure. See videos in [11].

Benefits of the Bent-Knee Position

One of the questions that came up when posing the 3-link *casteller* model was what is the advantage of the bent-knee position. To study this, we studied a system composed

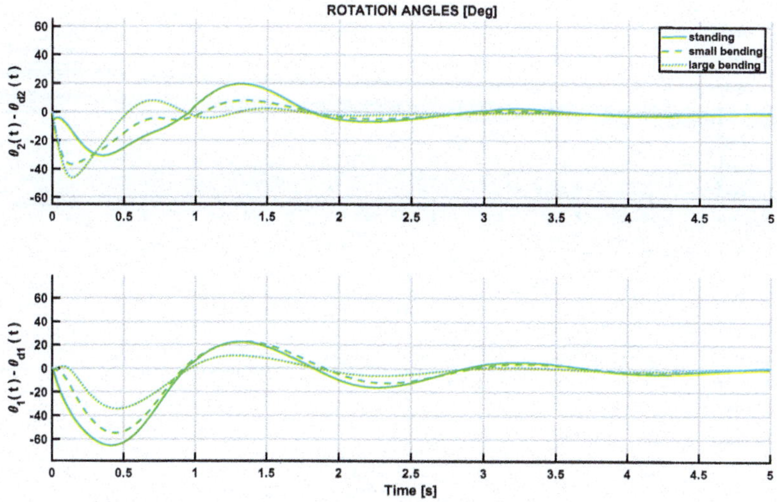

Fig. 1.7 Values for the difference between angles θ_1 and θ_2 and their corresponding equilibrium position as a function of time for the three bent-knee positions

by two links with an asymmetric joint. The equations for this system are the subset of equations for θ_1 and θ_2 in Sect. 1.3, setting $m_3 = 0$.

In this example we use the PD-like controller defined in (1.14) without any delay effects. The desired position is $\ddot{\boldsymbol{\theta}}_d = \dot{\boldsymbol{\theta}}_d = \mathbf{0}$ with a $\boldsymbol{\theta}_d$ to be specified. We impose that $\boldsymbol{\theta}_d$ must satisfy the condition that the centre of mass is aligned with the feet (similar to $d = 0$ in (1.15)), which leaves one degree of freedom in our problem. For the simulation we start at $\boldsymbol{\theta} = \boldsymbol{\theta}_d$ with initial angular velocities $\theta_1 = 0$ rad s^{-1} and $\theta_2 = 10$ rad s^{-1}. The rest of the model parameters are set to

$$
\begin{aligned}
l_1 &= 1.80, & c_1 &= 0.56, & m_1 &= 100, & J_1 &= 27.00, & K_{p1} &= 10000, & K_{d1} &= 100,\\
l_2 &= 2.25, & c_2 &= 0.56, & m_2 &= 125, & J_2 &= 52.73, & K_{p2} &= 12500, & K_{d2} &= 125,
\end{aligned}
$$

with $g = 9.81$.

To test the benefits of the bent-knee position, we run the simulation for three different desired positions: standing ($\theta_{d,1} = 0°$), small bending ($\theta_{d,1} = -10°$) and large bending ($\theta_{d,1} = -20°$). The desired position of θ_2, defined as $\theta_{d,2}$, is chosen to satisfy that the centre of mass is aligned with the feet.

The results of the simulations are shown in Figs. 1.7, 1.8 and 1.9. We observe that, even though the control torques are initially larger, the system returns faster to the equilibrium positions the more bent these are. We suspect that this is because, in those cases, the kinematic constraint does not act which makes it easier for *castellers* to recover balance.

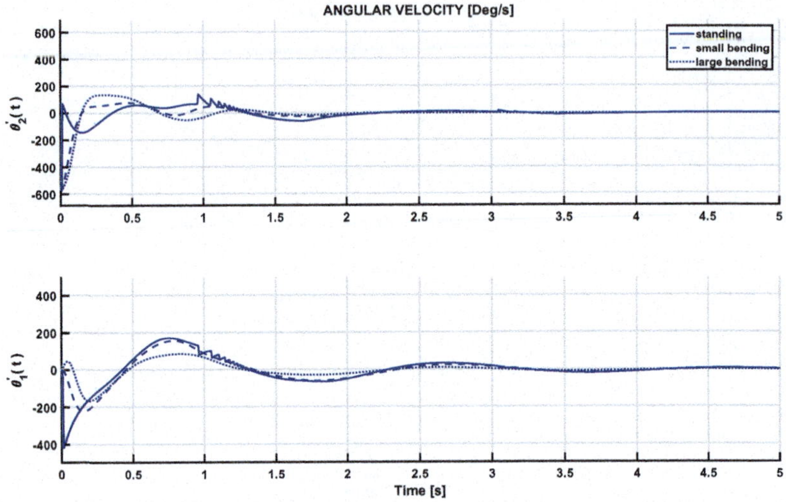

Fig. 1.8 Values for the angular velocities $\dot{\theta}_1$ and $\dot{\theta}_2$ as a function of time for the three bent-knee positions

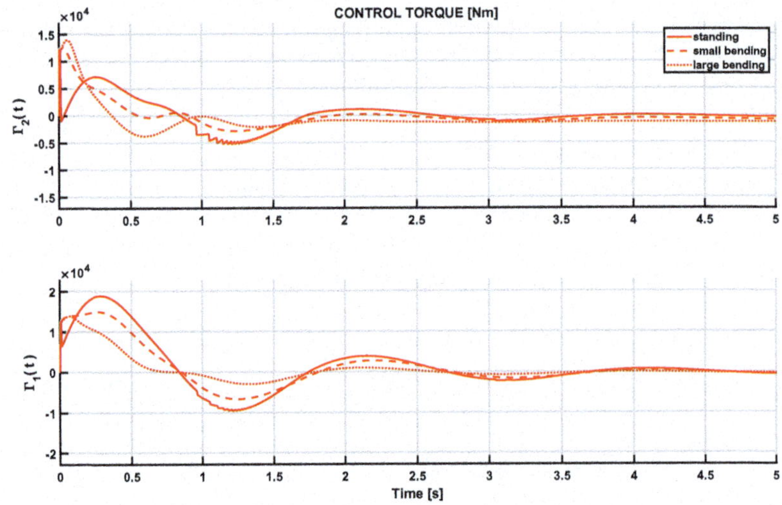

Fig. 1.9 Values for the torques Γ_1 and Γ_2 as a function of time for the three bent-knee positions

1.4 A Static Balance Point of View

The approach in this section is slightly different from the followed in the previous ones. We here consider a given and fixed configuration for a *pilar* and compute the torque distributions caused by the eccentricity of self-weighted loads in a non-perfectly vertical pillar. This is done considering a static balance of forces and moments, where the *castellers* are modelled again as a set of three links (see Fig. 1.10 and Sect. 1.3), statically interacting between them and, of course, the rest of the *castellers* of the *tronc*.

We start by recalling some notation given in the previous sections: each of the links (and corresponding joints) of a *pilar* is labelled i, j, where $i = 1 \ldots N$ represents the *casteller* (1 corresponding to the bottom one, N corresponding to the top one, see Sect. 1.2), and $j = 1, 2, 3$ represents the bars (and the joints) of each *casteller* (see Sect. 1.3). See Fig. 1.10 as example.

In the whole section, we are going to assume that no *castellers* are ascending or descending, that is no *castellers* are at the back of the other ones. The given configuration is parametrized by the bar rotations $\theta_{i,j}$ (measured counterclockwise from the vertical; the angle $\theta_{1,3}$ in Fig. 1.10 is negative). Finally, a set of biomechanical data will be also provided: $c_{i,j}, l_{i,j}, m_{i,j}$ represent the relative location of the mass center (i.e. $c_{i,j} = 0.5$ would mean a center of mass located at the midpoint of the bar), the length and the mass of the ij-bar.

In order to compute the torque distribution, we first need to compute some auxiliary variables. Using simple trigonometry it is easy to see that the horizontal displacement of the ij-joint, $d_{i,j}$ (with respect to the feet of *casteller* 1 and the usual sign convention, i.e rightward displacements are positive), can be computed as

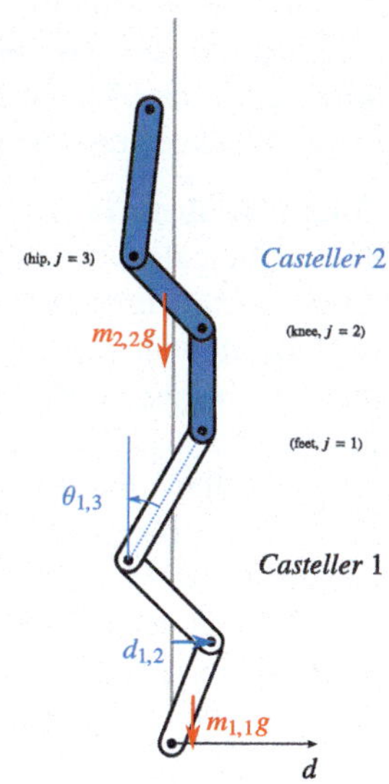

Fig. 1.10 Sketch of a static model configuration, that complements notation given in Fig. 1.5

$$
\begin{aligned}
d_{1,1} &= 0, \\
d_{i,1} &= d_{i-1,3} - l_{i-1,3} \sin(\theta_{i-1,3}), \quad \text{for } 1 < i \leq N, \\
d_{i,j} &= d_{i,j-1} - l_{i,j-1} \sin(\theta_{i,j-1}), \quad \text{for } 1 \leq i \leq N, \ j = 2, 3.
\end{aligned}
$$

From $d_{i,j}$ we can compute the horizontal displacement of the ij-bar mass center, $a_{i,j}$:

$$
\begin{aligned}
a_{i,j} &= (1 - c_{i,j})d_{i,j} + c_{i,j}\, d_{i,j+1}, \quad \text{for } 1 \leq i \leq N, \ j = 1, 2, \\
a_{i,3} &= (1 - c_{i,3})d_{i,3} + c_{i,3}\, d_{i+1,1}, \quad \text{for } 1 \leq i < N, \\
a_{N,3} &= d_{N,3} - c_{N,3}\, l_{N,3} \sin(\theta_{N,3}).
\end{aligned}
$$

and also the displacement of the i-*casteller* mass center is:

$$
d_i = \frac{\sum_{j=1}^{3} m_{i,j} a_{i,j}}{m_i}, \quad i = 1, \ldots, N, \tag{1.21}
$$

where $m_i = \sum_{j=1}^{3} m_{i,j}$ stands for the mass of i-*casteller*.

With all these ingredients, the torque $\Gamma_{i,j}$ supported at the ij-joint can be computed as

$$
\Gamma_{i,j} = \sum_{k=j}^{3} m_{i,k} g\left(a_{i,k} - d_{i,j}\right) + \sum_{k=i+1}^{N} m_k g\left(d_k - d_{i,j}\right), \quad i = 1, \ldots, N, \ j = 1, 2, 3, \tag{1.22}
$$

where g stands for the gravitational acceleration. The first summation term in the right hand side corresponds to the torque due to self-weight of *casteller* i above joint j, while the second one corresponds to the torque from the *castellers* above. We recall that we are assuming that no *castellers* are descending, that is, no *castellers* are at the back of the *casteller* i. With our sign conventions, a negative supported torque indicates that the pillar is leaning forward (to the right).

Quality indicators. The previous computations can be used to define three heuristic quality indicators for a *castell*, aimed at determining whether a given angle configuration is feasible to be maintained or, on the contrary, will collapse. All of them can be computed at each joint or averaged over each floor or the whole They complement the balance criterion given in Sect. 1.3, where no interactions with the rest of the floors are considered.

These indicators are the *mass center eccentricity* with respect to the vertical, the accumulated horizontal displacement or *folding index*, and the *suffering index*, whose objective is evaluating how much the joints are suffering.

The *mass center eccentricity* is the horizontal distance of the load above a given joint to the origin of coordinates (that is, the feet of the bottom-most *casteller*). We can compute this eccentricity in each floor or *casteller* (E_i) or for the whole *castell* (E_T):

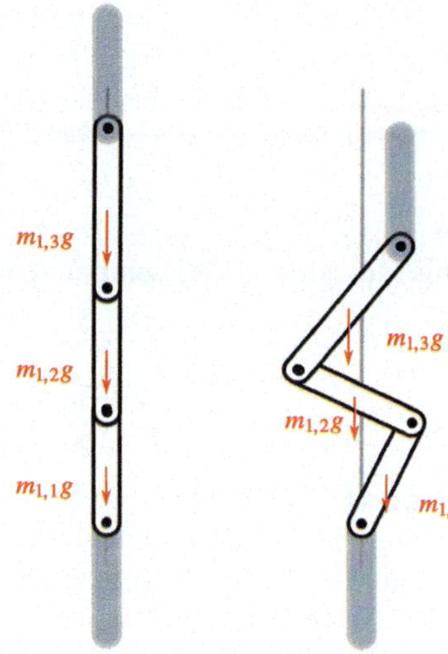

Fig. 1.11 Mass center eccentricity equals 0 for a perfectly vertical pillar (left) and a folded one (right). But folding index only equals 0 for the perfectly vertical pillar (left)

$$E_i = \frac{\Gamma_{i,1}}{\sum_{j=i}^{N} m_j\, g}, \quad E_T = \frac{\Gamma_{1,1}}{Mg} \tag{1.23}$$

(here $M = \sum_{i=1}^{N} m_i$ represents the total mass of the *castell*). It is a mechanical indicator of the average verticality of a *castell* in the following sense: a perfectly vertical pillar has $E_T = 0$, but a pillar with significant rotations $\theta_{i,j}$ may also have $E_T = 0$ if positive and negative rotations compensate each other (see Fig. 1.11). In this sense, it can be seen as an external indicator of the quality of a *castell*: if it is larger than a certain threshold, the *castell* bends to much with respect to the vertical and, hence, it is more likely to collapse.

The *folding index* is defined as the accumulated horizontal displacement. Again, it can be computed at each floor Δ_i or for the whole *castell* Δ_T:

$$\Delta_i = \sum_{j=1}^{3} |d_{i,j}|, \quad \Delta_T = \sum_{i=1}^{N} \Delta_i. \tag{1.24}$$

This index tells between structures with similar global deviation but its larger in those where the floors are more *folded*, that is, for the ones at which *castellers* are more bent. This is due to the fact that the absolute value precisely prevents for compensation when $d_{i,j}$ have different signs. Note that only a perfectly vertical pillar has $\Delta_T = 0$ (see Fig. 1.11). In this sense, we can think it also as an external, aesthetic and visual indicator of the quality of a *castell*. But we can also think it as a measure of the ability that a pillar has to react to perturbations, because a too large value of this index reveals a more fatiguing position for the involved *castellers* (and hence, less ability to react), but also a too small one may also mean that *castellers* are less able to react due to biomechanical reasons.

Finally, we define the *suffering index* as the ratio between the computed needed torque in each joint and the maximum torque this same joint can offer

$$S_{i,j} = \frac{\Gamma_{i,j}}{\Gamma_{i,j}^{max}}.$$

This indicator can be computed also in each floor, Γ_i^{rel}, or for the the whole *castell*, S_T,

$$S_i = \frac{\sum_{j=1}^{3} \Gamma_{i,j}}{\sum_{j=1}^{3} \Gamma_{i,j}^{max}}, \quad S_T = \frac{\sum_{i=1}^{N} \sum_{j=1}^{3} \Gamma_{i,j}}{\sum_{i=1}^{N} \sum_{j=1}^{3} \Gamma_{i,j}^{max}}.$$

It aims at quantifying how much the joints of the *castellers* are suffering in a certain position: if it is larger than a certain threshold, the *castell* will collapse, as that joint will not be physically able to maintain that torque. That would mean that with this indicator we could also get to know which particular joints are the ones that may have caused a *castell* to collapse. In this sense, we can think of the suffering index as an internal and mechanical indicator. It is worth noticing that the maximum allowed torques depend on biomechanical factors but also, of course, on the quality of a *casteller*, and that may decrease with time. In this sense it will be interesting to include in a future work the fatigue that the *castellers* are experiencing, as it may change the maximum moment allowed at certain joints.

Numerical results. To see how these quality indicators relate with the external and internal feelings of the quality of a *castell*, we evaluate them for two specific *pilars*, a successful one and one that ends up collapsing. Both of them are of the same type, *pilar de 8 amb folre i manilles (p8fm)*, which turns out to be the most difficult type of pillars that can be done.

The first one corresponds to the pillar that the *Colla Jove Xiquets de Tarragona (CJXT)* performed in the *Diada d'El Catllar* (2018) (see left picture of Fig. 1.12). And the one that ends up collapsing corresponds to the *Castellers de Vilafranca (CdV)*, and was performed in the *Diada d'Altafulla* (2014) (see pictures on the left column of Fig. 1.13). In both cases, we consider fixed frames of each pillar and measure the angle configuration of each one (one frame for the first pillar, but three for the second one, in order to compute all the variables at different stages of the failure). The corresponding angle configuration can be seen in Table 1.1. In a first approach, we estimate the following biomechanical parameters as

$$\begin{aligned} &c_{i,1} = 0.5, \quad c_{i,2} = 0.5, \quad c_{i,3} = 0.75 \\ &l_{i,1} = \frac{2}{7} L, \quad l_{i,2} = \frac{2}{7} L, \quad l_{i,3} = \frac{3}{7} L, \quad \text{with } L = 1 \\ &m_{i,1} = 0.1, \quad m_{i,2} = 0.2, \quad m_{i,3} = 0.7, \quad \text{with } M = 1 \end{aligned} \tag{1.25}$$

for all $i = 1, \ldots N$, and L standing for the height of a *casteller* (from feet to shoulders), that we take equal to 1 for simplicity. Observe that in this section, and again

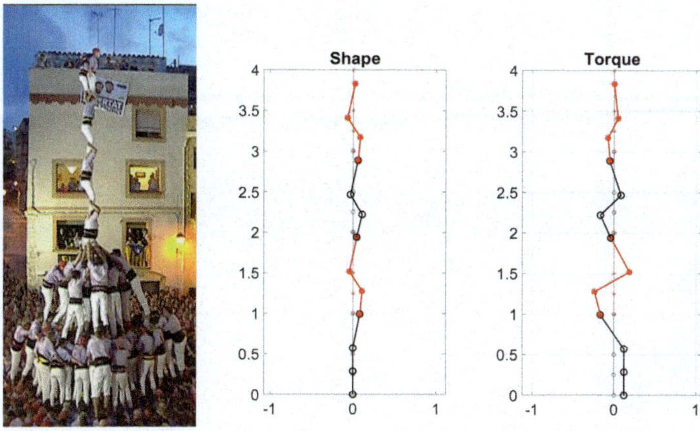

Fig. 1.12 Joint positions and received or supported torques in the *p8fm* of Colla Jove Xiquets de Tarragona (El Catllar, 2018) (frame taken from https://youtu.be/rCxT1rFLj9I, minute 3:35 aprox. Reproduced with permission from Xarxa Audiovisual Local, SL, www.laxarxa.cat). See the corresponding quality indicators on Table 1.2

for simplicity, we are assuming that the *pilar* is built with identical *castellers* in each floor. The appropriate values of these parameters are, of course, something that should be considered with more detail in a future work, as well as their heterogeneity among different *castellers*. Also, in these examples we are only considering the $N = 4$ *castellers* that go from levels four to seven, even though the *pilar* has eight levels in total (that is, $i = 1, 2, 3, 4$ correspond in this examples to the *quart*, *quint*, *sisè* and *acotxador*, respectively). We make this simplification because these are the most important floors in the *pilars* and the situations that we are trying to describe in these examples, at least in a first approach, and it suffices to give some insight in the approach here considered. But, of course, this analysis should be performed for the whole *pilar* in a future work.

By looking at these specific frames, it seems quite clear that the one in Fig. 1.12 corresponds to a *successful* pillar (it did not collapse, indeed), while the three in Fig. 1.13 show the evolution of a pillar that ends up collapsing (with the last frame in Fig. 1.13 showing the precise moment previous to this collapse). In both figures, the shape graphic (in the middle of these figures) is obtained by plotting the angle values given in Table 1.1. The torque graphic (at the right of these figures) corresponds to the torque received by each joint, computed according formula (1.22) and using the angles values given in Table 1.1 and the biomechanical parameters given in (1.25).

The shape and torque graphics show that, as expected, the configuration corresponding to the three frames of Fig. 1.13 is worse than the one corresponding to Fig. 1.12. And also that, among the three frames of Fig. 1.13, the torque at certain joints and *castellers* is increasing as the pillar is evolving and starting to collapse. We can see in all cases that a critical joint seems to be the one between the *quart*

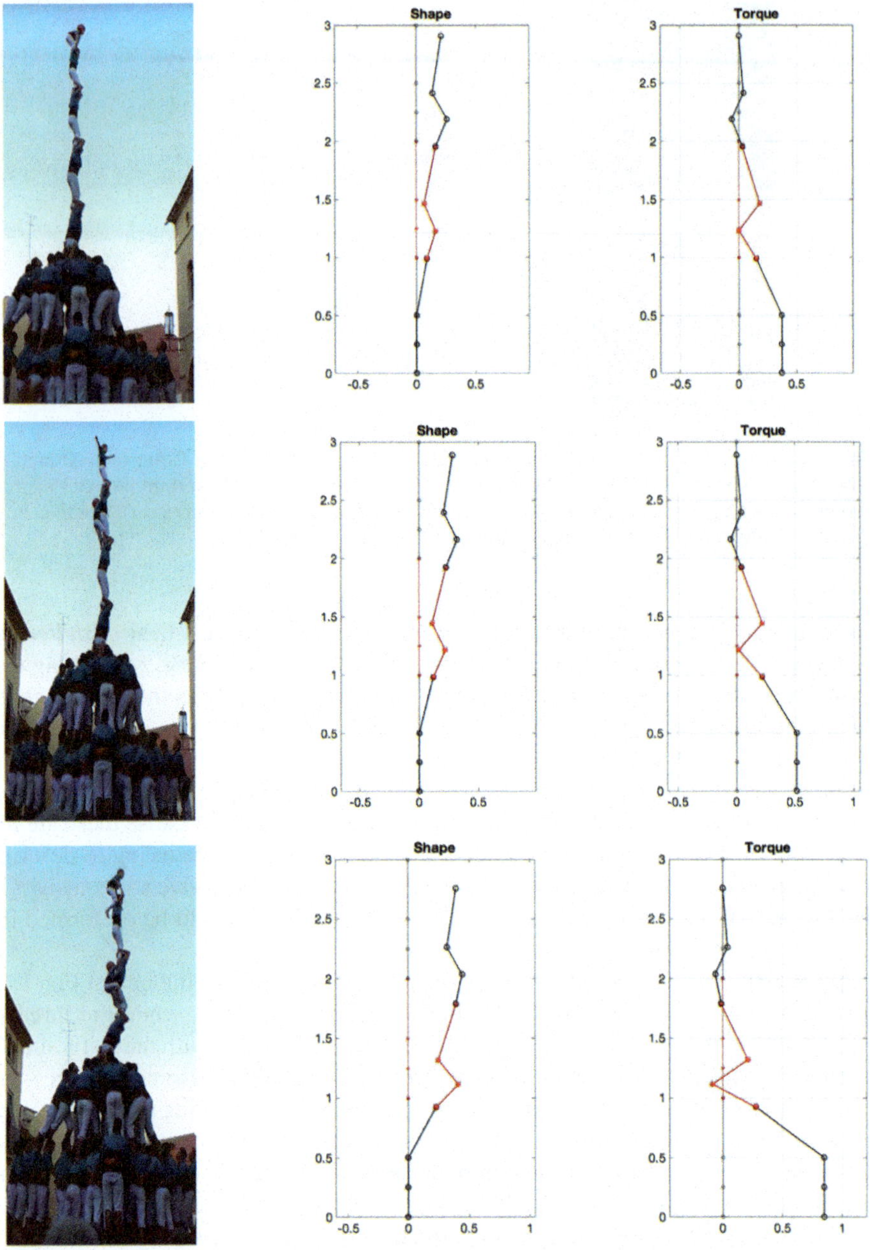

Fig. 1.13 Joint positions and received or supported torques in the *p8fm* of Castellers de Vilafranca (Altafulla, 2014) (frames taken from https://youtu.be/6o7Eaka7vcw, from minute 3:15). See the corresponding quality indicators on Table 1.2. Reproduced with permission from Josep Giménez

Table 1.1 Angle configuration (in degrees and with respect to the vertical) for the fixed configurations of the photos in the left of Figs. 1.12 and 1.13

Casteller	Joint angle	CJXdT	CdV (frame 1)	CdV (frame 2)	CdV (frame 3)
Quart ($i = 1$)	Feet (θ_{11})	0	0	0	0
	Knee (θ_{12})	0	0	0	0
	Hip (θ_{13})	−11	−12	−16	−32
Quint ($i = 2$)	Feet (θ_{21})	−5	−15	−21	−40
	Knee (θ_{22})	32	20	23	35
	Hip (θ_{23})	−12	−13	−16	−20
Sisè ($i = 3$)	Feet (θ_{31})	−13	−20	−19	−10
	Knee (θ_{32})	29	26	23	26
	Hip (θ_{33})	−12	−10	−10	−10
Acotxador ($i = 4$)	Feet (θ_{41})	−5	–	–	–
	Knee (θ_{42})	32	–	–	–
	Hip (θ_{43})	−12	–	–	–

and *quint castellers*, which is actually the joint where the pillar of Fig. 1.13 ends up breaking (see the link to the video given at the caption of Fig. 1.13).

With all these data we compute the first two quality indicators seen above (eccentricity and folding index, see (1.23) and (1.24), respectively). In future works it would be interesting to contact biomechanical experts in order to obtain the maximum torque allowed for an average feet, knee or hip joint, which is needed to compute the suffering index.

The results for the eccentricity and folding index can be seen in Table 1.2. Note that the computations related with the three frames do not take into account the effect of the *castellers* that are descending. (which, of course, is a very important disrupting factor for the stability of the structure and should be considered in a further work). This is also the reason for not having included the *acotxador* rotation in Table 1.1.

The values in Table 1.2 obtained in each case support the idea that these are indeed good indicators of the stability of a *pilar* and correlate positively with the external visual perception and the internal sensing of the quality of a *castell*. Indeed, it can be seen that all the indicators are lower in the case of the *good* pillar of CJXdT (Fig. 1.12) and higher as the pillar of CdV is getting worse (Fig. 1.13), both in each floor or for the whole *castell*. The last frame of this pillar (corresponding to the moment previous to the collapse) has the highest values of these quality indices. In addition, the eccentricity index corresponding to the *quart* of the pillar of CdV is also the highest one, and this also correlates with the fact that the breaking of the pillar is produced at the joint between the *quart* and *quint*.

These computations have been done using averaged, simulated or computed values of the involved parameters and data, to use the pillars of Figs. 1.12 and 1.13 as paradigmatic examples of the suggested approach, which seems to correlate with

Table 1.2 Summary of quality indicators for Figs. 1.12 and 1.13

Casteller	Joint angle	CJXdT	CdV (fm 1)	CdV (fm 2)	CdV (fm 3)
Mass center eccentricity	*Quart* ($i = 1$)	0.0310	0.1240	0.1718	0.2820
	Quint ($i = 2$)	−0.1365	−0.0156	−0.0096	−0.0908
	Sisè ($i = 3$)	−0.0638	−0.1300	−0.1889	−0.4075
	Acotxador ($i = 4$)	−0.1099	–	–	–
	TOTAL	**0.0310**	**0.1240**	**0.1718**	**0.2820**
Folding index	*Quart* ($i = 1$)	0.0000	0.0000	0.0000	0.0000
	Quint ($i = 2$)	0.2332	0.3175	0.4475	0.8848
	Sisè ($i = 3$)	0.1829	0.5554	0.7554	1.1544
	Acotxador ($i = 4$)	0.2106	–	–	–
	TOTAL	**0.6267**	**0.8729**	**1.2030**	**2.0391**

experience of *castells*. But, of course, and as a further work, it would be important to use proper biomechanical and physical data to compute the torques and the quality indices. As some of them may not be known, this future work would imply the design of experiments in order to obtain them.

1.5 Conclusions and Future Directions

Understanding which are the mechanics and mechanisms that allow a *castell* to be built and maintained is the first step for understanding the causes of its collapse and how and where do the *castellers* fall down once the structure is broken, and the main goal of this first work on the subject. Among all, the *pilar* is a type of *castell* that includes most of the important challenges to deal with in a first mechanical model. That is why we have focused our study in the *pilar* from three different points of view: static, dynamical and control model.

First, we have considered the dynamical problem thinking the *pilar* as N-link inverted pendula (each of the links corresponding to one *casteller*), when assuming passive (non reacting) *castellers*. Design of experiments in order to calibrate the model parameters should be designed. Also, a better understanding of the underlying dynamics of this model would be the first step to understand the physical reasons of a *pilar* collapse and, hence, to develop a first-failure model with impacts.

Second, assuming active response *castellers* would mean to consider the control problem. Also, to make this problem more realistic, we have considered a 3-link *casteller* model, where the movement of the joints of each *casteller* can be used as a control of the system. As a starting point for this model, we have considered the case of a single *casteller* with a passivity-based controller to achieve the desired position, where the interaction with the *castellers* of other floors not considered. The 3-link *casteller* has introduced the concept of internal joints, which represent the body articulations (in this case, knees and hip) and external ones (in this case, the joint between the feet of one *casteller* and the *casteller* below him). While the external ones can break (and this should be implemented in a future work), the internal ones cannot. Instead, they show an asymmetric behaviour as they cannot bend past a certain limit angle. In order to account for this, we have introduced a kinematic constraint into the model. We have considered a three conditions balance criterion for the 3-link *casteller*, which is related with the controller: align the centre of mass horizontally with the feet, align the shoulders horizontally with the feet and bring the centre of mass to a given height, which should represent a "comfortable" position for the *casteller*. We also introduced a delay-differential balance model which, even though we do not explore it in this work, could be included to account for the reaction time of a *casteller*.

The 3-link *casteller* model has arisen because, as we can observe in Fig. 1.4, *castellers* adopt a bent-knee position rather than standing perfectly vertical. This fact posed the question of what are the benefits of this position. In order to analyse this position, we have considered a 2-link model with an asymmetric joint and we ran simulations for three different desired positions (standing, small bending and large bending). We have found that, starting each simulation at the corresponding equilibrium position with the same initial velocity, the more bent the position is the faster it returned to equilibrium, even though it implied larger control torques at early times.

Again in this approach, the control parameters should be calibrated, first for an average configuration data (control to achieve a certain goal configuration) and, second, adapt them to dynamical data (control when the *pilar* is moving). In the first part, we should consider things such as biomechanics (what human body is able to do in order to achieve a certain position, which will give kinematic constraints to the problem) or the quality of a *casteller* (control parameters may be different depending on the ability of a *casteller*). In the second part, it would be important in future works to develop aspects such as the delay in the response to an external stimulus or the fatigue that the *castellers* are experiencing as the *castell* goes on (including them as a delay problem and, maybe, considering that values of the control parameters could change with time).

Further work on this control problem approach would include improving the control strategy by, for instance, implementing joint-coordinated control strategies for each *casteller* (control strategies taking into account all the joints of a *casteller* and not thinking them as independent) to decide its own control action, as well as incorporating in them the interaction with the rest of the floors. These joint torque control strategies should account, of course, for realistic dynamical feedback features of the human neuro-muscular system, including fatigue effects, as said above. Also, it would be very interesting (and challenging) the study of the static and dynamic stability of the realistic controlled model. One may think that several equilibria are possible, but not all of them with the same stability nor with the same cost to be achieved or maintained.

Finally, we have approached the mechanics in the static problem for the whole *pilar* considering again a 3-link *casteller* model: given a fixed configuration, we have obtained the torques at each joint of the *castellers* considering their own moments and those coming from the interaction with the other floors in the whole *pilar*. This has allowed us to compute three quality indicators that seem to correlate with the feasibility of a *castell*, and that complement the balance criterion used in the control problem above. The relation between feasible torques and angles should be investigated, as it may provide feasible zones for a successful *pilar*. It can also be used in order to validate an angle configuration given by the solutions of the dynamical and/or control problems at a fixed time.

In the previous discussions we have provided some simplified examples to give some insight on our suggested approaches. But, in order to have feasible results, proper and accurate biomechanical data should be used, such as maximum torques or angles allowed by particular joints in human body or lengths, mass densities and mass centers of human joints. Hence, we should look for the type of data that is already known in the scientific literature and design possible studies in order to collect the unknown one, taking into account biomechanical differences between *castellers* of different floors. Also, the design of a studies on how to compute and collect accurate angles of a configuration of a *pilar* at a fixed time (possibly involving computer vision techniques) should be considered.

Finally, future modelling of a collapsing *castell*, incorporating joint ruptures and impact phenomena, will ask for a more involved dynamical model and efficient numerical techniques. Also, in order to produce useful results, with respect to the end-problem addressed, a realistic dynamical computational tool should be produced, able to achieve a large number of numerical simulations, from which statistical results may be inferred concerning the most probable *castellers* "landing" locations, body positions and impact forces. In this work we have considered the *pillar* as a case study for the *castells* modelling, as it can be thought as a 1D-structure that has a 2D-movement in most of the situations. But we think the approaches and techniques described in this report in the case of a *pilar* are the first step to the modelling of a *castell* as a 3D complex structure. New challenging aspects such as, for instance, torsion within each floor should be then considered in future works.

Acknowledgements This problem was suggested by the Coordinadora de Colles Castelleres de Catalunya (CCCC). The authors would like to thank the CCCC, and specially its Scientific and Medical coordinator Dr. Dani Castillo, for this. Also, we would like to thank Dr. Jaume Rosset and Xavi Rius for their interesting and helpful discussions on the problem. This work is partially supported by the following grants. F. Brosa Planella has been supported by The Faraday Institution (faraday.ac.uk; EP/S003053/1), grant number FIRG003. A. Dòria-Cerezo has been partially supported by the *Generalitat de Catalunya* through the Project 2017 SGR 872. M. Pellicer is part of the Catalan Research group 2017 SGR 1392 and has been supported by the MINECO grant MTM2017-84214-C2-2-P (Spain).

Appendix A: Glossary of terms

In this section we introduce some of the terminology about *castellers* that has been used in this report (note that an 's' at the end of a word denotes the plural form). See also Fig. 1.1.

Each *castell* is named by the number of floors (N) and the number of *castellers* in each floor (M) and is usually abbreviated as MdN. For example, a *4d8* stands for a *castell* of 8 floors, with 4 people in each floor. When we have only one person per floor ($M = 1$) we name the *castell* as *pilar*. For instance, a *pd6* stands for a *castell* of 6 floors, with 1 person in each floor.

Acotxador: *casteller* (usually a child) that occupies the second-to-top level of a *castell.*
Ball de valencians: translated as Dance of the Valencians, they are believed to be the performances that originated the *castells* in the XVIII century.
Carregat/Descarregat: it is said a *castell* is *carregat* once the *enxaneta* reaches the top of it. The *castell* is considered to be *descarregat* once it is successfully disassembled.
Castell: human tower, of variable size and height, made of *castellers* standing on each other's shoulders. Typical from traditional festivals in Catalonia.
Casteller: person that takes part in the building of *castells*.
Colla: teams or groups that perform the *castells*, each of them of a different town, village or neighbourhood. They can be distinguished by the colour of their shirt.
Enxaneta: *casteller* (usually a child) at the top of a *castell.* Once he or she has reached and crossed the top of the structure, the *castell* is considered to be *carregat*.
Folre: Second basis that needs to be build on the *pinya* in higher *castells* to keep them stable.
Manilles: Third basis that needs to be build on the *manilles* in even higher *castells* to keep them stable.
Pilar: (or pillar) *castell* made of a single *casteller* at each level.
Pinya: group of people that hold the basis of a *castell.*
Pom de dalt: the top three levels of a castell (unless it is a *pilar*), with the *enxaneta* at the top one and the *acotxador* at the second one.
Quart: *casteller* that is in the fourth floor of a *castell.*
Quint: *casteller* that is in the fifth floor of a *castell.*
Sisè: *casteller* that is in the sixth floor of a *castell.*
Tronc: vertical part of a *castell.*

References

1. J.R. Chagdesa, S. Rietdykb, J.M. Haddadb, H.N. Zelaznikb, A. Raman, Dynamic stability of a human standing on a balance board. J. Biomechan. **46**(15), 2593–2602 (2013)
2. S. Kajita, Feedback control of inverted pendulums, in *Humanoid Robotics: A Reference* ed. by A. Goswami, P. Vadakkepat (Springer Nature, Dordrecht, 2019)

3. J.-Y. Kim, A.H. Vette, J.K. Mills, M.R. Popovic, Optimal combination of minimum degrees of freedom to be actuated in the lower limbs to facilitate arm-free paraplegic standing. J. Biomech. Eng. **129**, 838–847 (2007)
4. P. Kowalczyk, P. Glendinning, M. Brown, G. Medrano-Cerda, H. Dallali, J. Shapiro, Modelling human balance using switched systems with linear feedback control. J. R. Soc. Interface **9**, 234–245 (2012)
5. R. Ortega, J.A. Loría-Pérez, P.J. Nicklasson, H. Sira-Ramírez, *Passivity-Based Control of Euler-Lagrange Systems* (Springer, 1998)
6. J. Roset, E. Cardús, E. Honorato, J.M. Saló, *Risc de lesió en els castellers a partir del càlcul de l'energia potencial* (in Catalan). Apunts Medicina de l'esport 21–26 (1997)
7. J. Rosset, La ciència darrere els castells, in *Enciclopèdia castellera. Tècnica i ciència* (in Catalan), vol. 3 (Cossetània edicions, 2018), pp. 138–247
8. G. Seber, *A Matrix Handbook for Statisticians* (Wiley, Hoboken, 2008)
9. D.A. Winter, Human balance and posture control during standing and walking. Gait & Posture **3**, 193–214 (1995)
10. *El risc dels castells* (in Catalan). Coordinadora de Colles Castelleres de Catalunya (2016)
11. Video list with the simulations of the delay-differential balance model: https://www.youtube.com/playlist?list=PLS-qJKky1x-tC2cc7NgWLo2sF2bj-y9s9
12. Video of Human towers at UNESCO channel: https://youtu.be/-iSHfrmGdyo. From Departament de Cultura i Mitjans de Comunicació de la Generalitat de Catalunya (2009)
13. Web page of Coordinadora de Colles Castelleres de Catalunya (in English): http://www.cccc.cat/the-castells
14. Wikipedia page for *castells* (in English): https://en.wikipedia.org/wiki/Castell

Chapter 2
Safe Trajectory of a Piece Moved by a Robot

Thomas Babb, Ernest Benedito, Oliver Bond, Sandeep Kumar, Juan R. Pacha, and Joan Solà-Morales

Statement of the Problem

The company F.EE wants to determine the equations of the trajectory and the orientation of pieces that are moved by the action of a robot with 7 degrees of freedom, of which 6 correspond to rotations of the arms and 1 to longitudinal transfers. In many cases of interest, the piece is a curved metal sheet, and in others a three-dimensional object.

At the initial moment, the robot grabs the piece, which is at rest, at a certain point and orientation, and leaves it at rest at another given point and with a specific orientation, also, at the final instant. During its movement, the piece can suffer reversible and irreversible deformations caused by both mass forces (gravitational and inertial) and surface forces (those that the robot exerts to move it and those of the air) that act on it. The company wants that the resulting motion does not cause irreversible deformations of the piece and that the time between the initial and final moment does not exceed a certain threshold.

Certainly, the problem is complex and is rarely addressed in the scientific literature. Although there is abundant literature on the trajectory design of a piece moved by a robot with varying degrees of freedom, the deformations caused by mass forces or friction are generally not taken into account.

This is certainly a matter of practical interest to the manufacturing industry and it should also be of interest to the scientific community.

T. Babb · O. Bond
University of Oxford, Oxford, UK

E. Benedito · J. R. Pacha · J. Solà-Morales (✉)
Universitat Politècnica de Catalunya, Barcelona, Spain
e-mail: jc.sola-morales@upc.edu

S. Kumar
Basque Center for Applied Mathematics, Bilbao, Spain

M. Aguareles et al. (eds.), *Applications of Industrial Mathematics*,
RSME Springer Series 8, https://doi.org/10.1007/978-3-031-32130-6_2

Fig. 2.1 Top left: Metal piece moved by the arm of the robot; top right: Metal piece as a part of a car; bottom: A real-life scenario at the manufacturing site

Due to its complexity, we understand that the problem cannot be solved in as limited a time as the 5 days of the ESGI meeting, but in this time, the first steps can be taken to solve it in the future (Fig. 2.1).

Therefore, we propose to limit the objective of the problem to be addressed during the ESGI days and propose the following tasks:

- Perform a literature search on research in this area. Search also literature on geometry paths that avoid non-progressive changes in accelerations. Explore available software to calculate forces and deformations due to mass and surface forces of moving bodies.
- Define a simple case of the problem: for example, the movement of a "longitudinal" piece, held at one end, to a body that can move in a direction perpendicular to that of the force of gravity, and can rotate the piece in the plane defined by the direction of motion of the body and that of the force of gravity.
- Write the equations of movement of the body-fluid system, in the most general case possible, by determining the stresses and deformations suffered by a body due to the forces of mass and friction acting on it when being moved.
- Implement a simplified problem model in any of the programs explored.
- Propose a draft of one or more possible research projects for a more complete study of the topic.

2.1 Introduction

The company F.EE (www.fee.de/en.html) is an international company supplying automation technologies for industry, mostly in the automotive sector. The company is interested in determining the equations of the trajectories and the orientations of pieces that are moved by the action of a robot with seven degrees of freedom, of which six correspond to rotations of the arms and one to longitudinal transfers. The piece can be a curved metal sheet or a three-dimensional object.

Initially, the piece is at a certain point/location, orientation, and rest position, then the robot grabs and leaves it to another given point/location, orientation, and rest position corresponding to the later time instant. During its movement, the piece can suffer reversible and irreversible deformations due body and surface forces. The body forces are due to translation and rotations (depending on accelerations and angular velocity of the body) while surface forces are due to air drag (depending on the velocity of the body). It is desirable that the resulting motion does not result in irreversible deformations on the piece and that the time between the initial and final moments does not exceed a certain threshold. The actual movement of a piece is a consequence of the movements of the robot. This in turn depends on the design of the trajectory to be followed by the piece, as well as on the robot's ability to faithfully follow the desired trajectory. Any solution to the problem must ensure that the desired trajectory does not cause irreversible deformation but also that the robot can actually follow the suggested trajectory. The former depends on the desired velocity, acceleration, angular velocity of the piece, and the latter depends on the so-called *jerk*, or, the first-derivative of the accelerations.

Certainly, the problem is complex and is rarely addressed in the scientific literature. There is abundant literature on the trajectory design of a piece moved by a robot with varying degrees of freedom [1, 2] but the deformations caused by mass forces or friction are generally not taken into account. This aspect is certainly important for the manufacturing industry for various practical reasons, thus, should be discussed in the scientific community as well.

As already pointed out that due to the complexity, in a five days period it is only expected that the mathematical framework of the problem may be laid out, which can be later adapted and solved. Therefore, during the ESGI days, we have restricted the objectives of the problem to the following tasks:

- Perform a literature search in this area, including a search of numerical codes to solve these kinds of problems.
- Write the equations of motion of the body-fluid system, in the most general case possible, by determining the stresses and deformations suffered by a body due to the forces of mass and friction acting on it when being moved by another.
- Together with the possible use of commercial software, discuss numerical algorithms to solve these equations.
- Define a very simple case of the problem. In this case we chose to study the movement of a longitudinal piece, held at one end by a body that can move in a direction perpendicular to that of the force of gravity, and can rotate the piece in

the plane defined by the direction of motion of the body and that of the force of gravity.

The results of the first point will be described in Sect. 2.5 below, together with conclusions about possible future work, and the next three points will be described in Sects. 2.2, 2.3, and 2.4.

We want to point out that we have devoted our efforts to the so-called *direct* problem, rather that the more difficult, but perhaps more important in practice, *inverse* problem. Let us explain the difference: In the *direct* problem, one assumes that the trajectory of the arm and its rotations are known and given, and the problem is to calculate deformations and stresses of the piece. In the *inverse* problem, the goal is to determine trajectories and rotations such that the induced deformations and stresses satisfy the restriction requirements given by the nature of the pieces, to avoid permanent deformations. Even if one considers that only the inverse problem is of practical interest, it is clear that the solution of the inverse problem requires the solution of many direct problems, to achieve the optimal candidate.

2.2 The Equations

Before explaining the equations and boundary conditions specific to plate theory, we provide a brief introduction to the governing equations of solid mechanics. This explanation is a condensed version of that found in [3], which provides a much more rigorous introduction to the field.

Introduction to Solid Mechanics

Suppose that any point inside an elastic object $\mathcal{P} \subset \mathbb{R}^3$ has an initial *Lagrangian* coordinate $\vec{X} \in \mathcal{P}$, which is fixed in the material. At time $t > 0$, this coordinate changes with elastic deformations of the object, into an *Eulerian* coordinate $\vec{x}(\vec{X}, t)$, where the coordinate $\vec{x}$ remains fixed in space. From these, we can construct the *displacement field* $\vec{u}(\vec{x}, t) := \vec{x}(\vec{X}, t) - \vec{X}$, which in part is found by formulating an equation relating body and surface forces, and using conservation of momentum to obtain a partial differential equation known as the *Navier equation* in which $\vec{u}(\vec{x}, t)$ is the dependent variable.

Consider an arbitrary volume V contained within the elastic material. This small volume experiences body forces (e.g. due to gravity) and surface forces. In the case where the density ρ is uniform throughout the material, one can write down expressions for both of these effects, as well as the rate of change of momentum (where momentum is the mass of an object multiplied by its velocity). In this case, the ith component of the displacement field, u_i, satisfies

$$\underbrace{\frac{\mathrm{d}}{\mathrm{d}t}\iiint_V \frac{\partial u_i}{\partial t}\rho \mathrm{d}\mathcal{P}}_{\text{rate of change of momentum}} = \underbrace{\iiint_V g_i \rho \mathrm{d}\mathcal{P}}_{\text{total body force}} + \underbrace{\iint_{\partial V} \sigma_{ij} n_j \mathrm{d}S}_{\text{total surface force}},$$

where g_i is the i-th component of the body force, and σ_{ij} is the (symmetric) stress tensor, which tells us the ith component of the force per unit area acting at a point on the surface with outward normal n_j. By taking the time derivative inside the first term (using the fact that the density is constant) and applying the divergence theorem on the final term, we can use the fact that V is arbitrary and the assumption that each integrand is continuous to obtain *Cauchy's momentum equation*

$$\rho \frac{\partial^2 u_i}{\partial t^2} = \rho g_i + \frac{\partial \sigma_{ij}}{\partial x_j}. \tag{2.1}$$

To be able to determine the displacement field relies on us knowing the stress tensor σ_{ij}. For this there is a constitutive relation, known as *Hooke's law*, which postulates a linear relationship between the stress and another quantity called the *strain*, a dimensionless quantity signifying the elastic extension of an object relative to its original state. To identify the exact form of the strain tensor, we can consider two particles with positions $\vec{X}$ and $\vec{X} + \delta\vec{X}$, displaced to $\vec{x} = \vec{X} + \vec{u}(\vec{X}, t)$ and $\vec{x} + \delta\vec{x} = \vec{X} + \delta\vec{X} + \vec{u}(\vec{X} + \delta\vec{X}, t)$. Taylor's theorem can be used to show that (upon neglecting quadratic terms)

$$|\delta\vec{x}|^2 = \left|\delta\vec{X} + (\delta\vec{X} \cdot \nabla_{\vec{X}})\vec{u}(\vec{X}, t)\right|^2,$$

and therefore that

$$|\delta\vec{x}|^2 - \left|\delta\vec{X}\right|^2 = 2\varepsilon_{ij}\delta X_i \delta X_j,$$

where the strain tensor ε_{ij} is given by

$$\varepsilon_{ij} = \frac{1}{2}\left(\frac{\partial u_i}{\partial X_j} + \frac{\partial u_j}{\partial X_i} + \frac{\partial u_k}{\partial X_i}\frac{\partial u_k}{\partial X_j}\right) \sim \frac{1}{2}\left(\frac{\partial u_i}{\partial X_j} + \frac{\partial u_j}{\partial X_i}\right). \tag{2.2}$$

If the stress and strain are scalars, then Hooke's law simply states that $\sigma = E\varepsilon$, where E is a constant known as the *Young modulus*. However, in this context, σ_{ij} and ε_{ij} are both rank-2 tensors, so a linear relation between them must involve a rank-4 tensor, so that

$$\sigma_{ij} = C_{ijkl}\varepsilon_{kl}, \qquad i, j, k, l = 1, 2, 3.$$

Here, C_{ijkl} has 81 entries in total. However, by considering the stress acting at the surface of the volume V (by considering a pillbox-type argument), one can show that the rank-2 tensors are symmetric (i.e. $\sigma_{ij} = \sigma_{ji}$ and $\varepsilon_{ij} = \varepsilon_{ji}$). Under further modelling assumptions, such as homogeneity and isotropy of the material, one can derive a stress-strain relation of the form

$$\sigma_{ij} = \lambda \varepsilon_{kk} \delta_{ij} + 2\mu \varepsilon_{ij}, \tag{2.3}$$

where λ is the bulk modulus and μ is the shear modulus of the material. Both of these constants are collectively called the *Lamé constants* and they are material properties which indicate a material's tendency to withstand deformation. Upon substituting (2.3) and (2.2) into (2.1), one obtains the *Navier equation*, written in vector form as

$$\rho \frac{\partial^2 \vec{u}}{\partial t^2} = \rho \vec{g} + (\lambda + \mu) \nabla (\nabla \cdot \vec{u}) + \mu \nabla^2 \vec{u}. \tag{2.4}$$

By imposing suitable boundary conditions, one can solve equation (2.4) for the displacement field $\vec{u}(\vec{X}, t)$. The usual techniques of applied mathematics, such as asymptotic analysis and numerical methods, can be used to solve this equation, although numerical methods will be much more appropriate for arbitrary geometries (such as those from CAD files provided by component manufacturers).

Classical Plate Theory

A considerable simplification in our situation is that many components moved around by a robot are three-dimensional but are thin; for example, a car door. Although many people think of this type of component as strong, namely due to being made of metal, thin metal sheets are prone to elastic displacements which could exceed an elastic limit, deforming plastically and then being unsuitable for use in any further manufacturing (and thus needing to be recycled). Thankfully, a theory has been developed to tackle the case where the component is large and thin. This is called *Kirschoff-Love theory* (or *classical plate theory)*, where in-plane displacements are disregarded, and that displacements normal to the plane are considered relative to a *mid-plane*: a surface equidistant between the top and bottom face of the plane. This idea is illustrated in Fig. 2.2, although we emphasise that the plate need not be uniform everywhere since the coordinates are local to the plate. We guide the reader through the basics of this theory; and refer them to [4] for a much more detailed explanation. In the specific case shown in Fig. 2.2, an educated guess for the displacement field is given. This is of the form

$$u_i(x, y, z, t) = \sum_j (z)^j \phi_i^{(j)}(x, y, t),$$

where $\phi^{(j)}$ are functions chosen so that the *principle of virtual displacements* is satisfied and z is the coordinate in the direction of thickness. A *virtual displacement* of an elastic system is an infinitesimal displacement which can one might suspect to arise based on the current configuration of forces and if it was perturbed slightly. For example, if an elastic beam of length L is fixed at a wall at $x = 0$ and being pulled away by a force F being applied at its free end, then the beam can be thought of

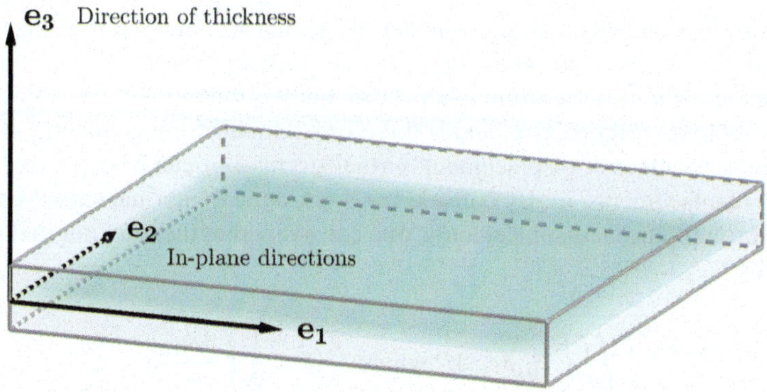

Fig. 2.2 A simple illustration of the coordinate system used in Kirchoff-Love plate theory for a (flat) plate

as having a "virtual" displacement of $\delta u(L)$ at its free end arising from a "virtual" force δF. The idea of the displacements being "virtual" is so-called because they are fictional and are unrelated to the displacements and actual loads on the system.

Virtual Work and Virtual Displacement

The idea of "virtual work" arises in two possible ways. We firstly recall that the work done by a force is the product of its projection (dot product) onto its direction of displacement, and its magnitude (or, put simply, "work is force times distance"). Virtual work can be considered to arise either from (a) an *actual* force moving through a *virtual displacement*, or (b) a *virtual force* moving through an *actual* displacement. If a lowercase delta before a variable denotes an infinitesimal virtual change in that variable, then the virtual work is given by

$$\delta W = \iiint_{\mathcal{P}} \vec{F} \cdot \delta \vec{u} \mathrm{d}\mathcal{P}.$$

Virtual work can be classified as either *external* or *internal*. Forces applied by external sources will do work when moving through virtual displacements, and to this effect one can write down the *total external virtual work*. When an elastic body is subjected to body forces of $\vec{f}$ per unit volume and surface tractions $\vec{T}$ per unit surface area (where Γ_σ is the subset of $\partial\mathcal{P}$ on which stresses are specified), upon moving through virtual displacements $\delta\vec{u}$ it will do a total amount of virtual work δV due to applied forces. This is given by

$$\delta V = -\left(\iiint_{\mathcal{P}} \vec{f} \cdot \delta \vec{u} \mathrm{d}\mathcal{P} + \iint_{\Gamma_\sigma} \vec{T} \cdot \delta \vec{u} \mathrm{d}S\right). \tag{2.5}$$

A deforming elastic object will also undergo internal stresses and then the work can be obtained in terms of the total strain inside it rather than due to specific vector forces. In simple terms, the strain of an elastic body is the ratio of the extension of an elastic body to its original size, and is therefore a dimensionless quantity, and since it involves derivatives of displacement, virtual strains $\delta\varepsilon_{ij}$ can be expressed in terms of virtual displacements δu. By considering work done by normal stresses and shear stresses through virtual displacements, one can show that the total internal work δU is given by

$$\delta U = \sum_{i=1}^{3}\sum_{j=1}^{3}\iiint_{\mathcal{P}} \sigma_{ij}\delta\varepsilon_{ij}\mathrm{d}\mathcal{P} \text{ where } \delta\varepsilon_{ij} = \frac{1}{2}\left(\frac{\partial \delta u_i}{\partial x_j} + \frac{\partial \delta u_j}{\partial x_i}\right). \tag{2.6}$$

At this point, we can concisely state the principle of virtual displacements, i.e. that a continuous body in equilibrium will have a total virtual work by actual forces through virtual displacements of zero. Put simply, $\delta V + \delta U = 0$; as each of these terms are integrals, the problem then becomes about minimising these integrals (through a *weak formulation*) rather than solving for an unknown function with a differential equation (through a *strong formulation*). More specifically of interest to us, the dynamical version of the principle of virtual displacements is

$$\int_0^T (\delta U + \delta V - \delta K)\,\mathrm{d}t = 0, \tag{2.7}$$

where δK is the virtual kinetic energy of the system. This method of deriving governing equations for an elastic body is an attractive alternative to Eq. (2.4) because (a) it does not need to rely on the use of constitutive laws upfront (e.g. Hooke's law), and (b) the elasticity problem can be posed in the form of a variational problem, thereby making it amenable to both analytical treatments (e.g. calculus of variations, Euler-Lagrange equations, Hamilton's principle) and contemporary numerical methods such as finite element methods (FEM). How one might approach this problem using FEM is to be discussed later.

Deriving the Governing Equation

We are interested in *Kirchoff's hypothesis* which provides the elastic displacement field a priori as

$$\begin{aligned}
u(x, y, z, t) &= u_0(x, y, t) - z\frac{\partial w_0}{\partial x}, \\
v(x, y, z, t) &= v_0(x, y, t) - z\frac{\partial w_0}{\partial y}, \\
w(x, y, z, t) &= w_0(x, y, t),
\end{aligned}$$

where u_0, v_0 and w_0 are displacements from the mid-plane in the x, y and z-directions respectively. In particular, we can set $u_0 \equiv 0$ and $v_0 \equiv 0$, due to neglecting in-plane displacements. Namely, we are interested in the domain $[-w, w] \times \mathcal{F}$, where $\mathcal{F}$ represents the cross-section of the plate at $z = 0$. Substituting the above form of the displacement field into the linearised strains (2.2), neglecting in-plane displacements, it is straightforward to show that the strain tensor is given by

$$\begin{aligned}
\epsilon_{11} &= \frac{1}{2}\left(\frac{\partial w_0}{\partial x}\right)^2 - z\frac{\partial^2 w_0}{\partial x^2} \\
\epsilon_{22} &= \frac{1}{2}\left(\frac{\partial w_0}{\partial y}\right)^2 - z\frac{\partial^2 w_0}{\partial y^2} \\
\epsilon_{12} &= \frac{1}{2}\left(\frac{\partial w_0}{\partial x}\frac{\partial w_0}{\partial y} - 2z\frac{\partial^2 w_0}{\partial x \partial y}\right) \\
\epsilon_{13} &= \frac{1}{2}\left(-\frac{\partial w_0}{\partial x} + \frac{\partial w_0}{\partial x}\right) = 0 \\
\epsilon_{23} &= \frac{1}{2}\left(-\frac{\partial w_0}{\partial y} + \frac{\partial w_0}{\partial y}\right) = 0 \\
\epsilon_{33} &= 0,
\end{aligned} \tag{2.8}$$

where we have given six entries instead of nine, owing to the fact that the strain tensor is symmetric (so that $\epsilon_{ij} = \epsilon_{ji}$). This puts us in a position to calculate the internal virtual work δU, given by

$$\begin{aligned}
\delta U &= \int_{-w}^{w}\int_{\mathcal{F}} (\sigma_{11}\delta\epsilon_{11} + 2\sigma_{12}\delta\epsilon_{12} + \sigma_{22}\delta\epsilon_{22})\, \mathrm{d}\mathcal{F}\,\mathrm{d}z \\
&= -\int_{-w}^{w}\int_{\mathcal{F}} \left(\sigma_{11}z\frac{\partial^2 \delta w_0}{\partial x^2} + 2\sigma_{12}z\frac{\partial^2 \delta w_0}{\partial x \partial y} + \sigma_{22}z\frac{\partial^2 \delta w_0}{\partial y^2}\right) \mathrm{d}\mathcal{F}\,\mathrm{d}z + \text{h.o.t.} \\
&= -\int_{\mathcal{F}} \left(M_{11}\frac{\partial^2 \delta w_0}{\partial x^2} + 2M_{12}\frac{\partial^2 \delta w_0}{\partial x \partial y} + M_{22}\frac{\partial^2 \delta w_0}{\partial y^2}\right) \mathrm{d}\mathcal{F} + \text{h.o.t.}
\end{aligned}$$

where "h.o.t." is an abbreviation for "higher-order terms"; terms which are quadratic in the partial derivatives of δw_0 and can therefore be neglected (although this is no longer appropriate if von Karman strains are considered). We also have

$$M_{ij} = \int_{-w}^{w} \sigma_{ij} z\,\mathrm{d}z, \tag{2.9}$$

are the *stress moment resultants*. We emphasise that in general, there will also be *stress resultants*

$$N_{ij} = \int_{-w}^{w} \sigma_{ij}\,\mathrm{d}z,$$

but neglecting the terms which are quadratic in the derivatives of w_0 leads to them being absent from the expression for the internal virtual work.

Meanwhile, the external virtual work is given by

$$\delta V = -\int_{-w}^{w}\int_{\mathcal{F}} \vec{f}\cdot\delta\vec{u}\mathrm{d}\mathcal{F}\mathrm{d}z - \int_{-w}^{w}\int_{\partial\mathcal{F}} \vec{T}\cdot\delta\vec{u}\mathrm{d}S\mathrm{d}z \tag{2.10}$$

$$= -\int_{-w}^{w}\int_{\mathcal{F}} (q - kw_0)\,\delta w_0 \mathrm{d}x\mathrm{d}y. \tag{2.11}$$

The second term in Eq. (2.10) disappears because the traction force, $T_i = \sigma_{ij}n_j$ where $\vec{n}$ is the normal to $\partial\mathcal{F}$ (a closed curve in the (x, y)-plane), is perpendicular to the z-direction in which the displacement is solely assumed to take place. In Eq. (2.11), $q(x, y)$ is the net load on $\mathcal{F}$, and the $-kw_0$ contribution arises from Hooke's law, where k is the stiffness constant of the plate material. The internal kinetic energy δK is given by

$$\begin{aligned}\delta K =& \int_{\mathcal{P}}\int_{-w}^{w} \rho\left(\dot{u}\delta\dot{u} + \dot{w}\delta\dot{w} + \dot{w}\delta\dot{w}\right)\mathrm{d}z\mathrm{d}x\mathrm{d}y \\ =& - I_0\int_{\mathcal{P}}\left(\dot{u}_0\delta\dot{u}_0 + \dot{w}_0\delta\dot{w}_0 + \dot{w}_0\delta\dot{w}_0\right)\mathrm{d}x\mathrm{d}y \\ &- I_2\int_{\mathcal{P}}\left(\frac{\partial\dot{w}_0}{\partial x}\frac{\partial\delta\dot{w}_0}{\partial x} + \frac{\partial\dot{w}_0}{\partial y}\frac{\partial\delta\dot{w}_0}{\partial y}\right)\mathrm{d}x\mathrm{d}y\end{aligned} \tag{2.12}$$

where the moments of inertia are given by

$$I_k = \int_{-w}^{w} z^k\rho\mathrm{d}z \text{ where } k = 0, 1, 2.$$

We do not present the full derivation of the governing equations here due to the amount of algebra involved, but briefly describe how it is done (see pages 103-105 of [4] for a full derivation). By substituting results (2.6), (2.11) and (2.12) into (2.7), making use of the virtual strains, employing techniques from the calculus of variations and setting the coefficient of δw_0 to zero, one can show that

$$\frac{\partial^2 M_{11}}{\partial x^2} + 2\frac{\partial^2 M_{12}}{\partial x\partial y} + \frac{\partial^2 M_{22}}{\partial y^2} - kw_0 + q = I_0\frac{\partial^2 w_0}{\partial t^2} - I_2\frac{\partial^2}{\partial t^2}\left(\frac{\partial^2 w_0}{\partial x^2} + \frac{\partial^2 w_0}{\partial y^2}\right).$$

Finally, it is desirable to express this governing equation in terms of displacements rather than moments. For a homogeneous, isotropic plate, a constitutive law tells us that

$$\begin{pmatrix}\sigma_{11}\\ \sigma_{22}\\ \sigma_{12}\end{pmatrix} = \frac{E}{1-\nu^2}\begin{pmatrix}1 & \nu & 0\\ \nu & 1 & 0\\ 0 & 0 & 1-\nu\end{pmatrix}\begin{pmatrix}\epsilon_{11}\\ \epsilon_{22}\\ \epsilon_{12}\end{pmatrix}, \tag{2.13}$$

where E is the Young modulus of the material, and ν is the Poisson ratio. Combining the constitutive law (2.13) with (2.9) and the strain-displacement relations (2.8), one obtains

$$D\nabla^2\nabla^2 w_0 = -q(x, y, t) - 2\rho h \frac{\partial^2 w_0}{\partial t^2}, \tag{2.14}$$

where the *bending stiffness*, D, is given by

$$D = \frac{2h^3 E}{3(1-\nu^2)}, \tag{2.15}$$

which will be loosely referred to as the "*Kirchoff-Love equation*".

Initial and Boundary Conditions

Along with the equation of motion (2.14), several boundary conditions based on likely physics in a factory setting need to be imposed. Let $C \subset \partial\mathcal{F}$ be the part of the large face boundary which is clamped by the robot, then the plate

1. is *initially undeformed*, i.e.,

$$w_0(x, y, t = 0) = 0,$$

2. is *initially stationary*, i.e.,

$$\frac{\partial w_0}{\partial t}(x, y, t = 0) = 0,$$

3. *does not deform* where it is clamped, i.e.,

$$w_0(x, y, t) = 0, \;\; x, y \in C;$$

4. and *no bending moments or loads* at the free boundary:

$$\nabla^2 w_0(x, y, t) = 0, \;\; \frac{\partial}{\partial n}(\nabla^2 w_0(x, y, t)) = 0, \;\; (x, y) \in \partial\mathcal{F}\backslash C.$$

Changing the Frame

As it stands, it suffices to solve equation (2.14) for the transverse elastic displacements of the plate, provided that the plate is not being moved externally. Considering how the plate is being moved around by a robot, this is not satisfactory; if the plate is

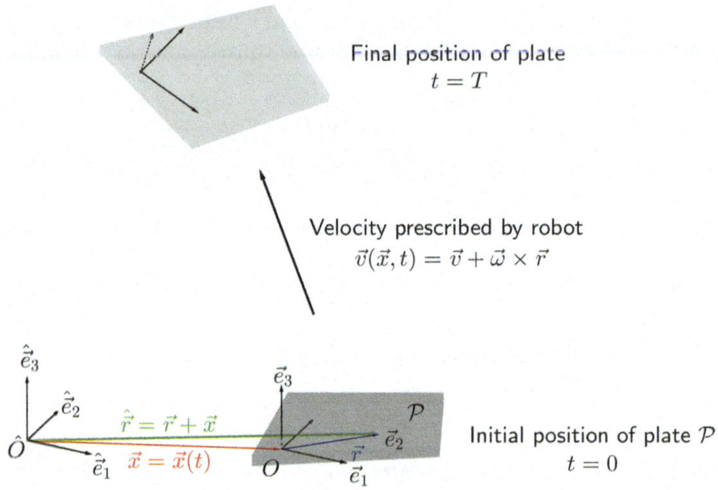

Fig. 2.3 An illustration of plate motion in two frames of reference

being moved in a non-inertial frame, it will experience "fictitious" forces, namely the Euler, Coriolis and centrifugal forces. These forces are so-called because they arise from a change of frame rather than a physical mechanism. They are mathematical in nature and should therefore be incorporated within the external load $q(x, y, t)$ as it appears in Eq. (2.14). The mathematical techniques for changing frames are standard and can be found in classical mechanics texts such as [5].

We use $\hat{\mathcal{S}}$ to denote an *inertial* frame which remains fixed over time, with origin $\hat{O}$ which is fixed in space, and orthonormal basis $\{\hat{\vec{e}}_1, \hat{\vec{e}}_2, \hat{\vec{e}}_3\}$. We use $\mathcal{S}$ to denote the *non-inertial* frame, which is fixed local to the plate and whose origin O moves with angular velocity $\vec{\omega}$ relative to $\hat{O}$. The relationship between the two coordinate systems is illustrated in Fig. 2.3.

F.EE have complete control over the rotational movement of the robotic arm, as well as translational motion along a straight line. In practice, the rotational motion of the robot motion may be prescribed by a *rotation matrix* $\mathcal{R}_{ij}(t)$ which acts as a time-dependent linear transformation from $\hat{\mathcal{S}}$ to $\mathcal{S}$ and may be expressed in terms of Euler angles. If $\hat{\mathrm{D}}$ and D denote time derivatives in the inertial and non-inertial frames respectively, and $\vec{r}$ is the position vector of a point in $\mathcal{S}$ fixed on the plate relative to O, then the Coriolis formula is given by

$$\hat{\mathrm{D}}\vec{r} = \mathrm{D}\vec{r} + \vec{\omega} \times \vec{r},$$

where $\vec{\omega}$ is the *angular velocity* vector whose entries satisfy

$$\mathcal{R}_{ij}\dot{\mathcal{R}}_{ji} = \sum_{k=1}^{3} \epsilon_{ijk}\omega_k. \tag{2.16}$$

Applying the Coriolis formula twice can be used to obtain an expression for the acceleration $\hat{\vec{a}}$ in the inertial frame in terms of the acceleration $\vec{a}$ in the non-inertial frames:

$$\hat{\vec{a}} = \vec{a} + (\mathrm{D}\vec{\omega}) \times \vec{r} + 2\vec{\omega} \times \mathrm{D}\vec{r} + \vec{\omega} \times (\vec{\omega} \times \vec{r}) + \vec{A}, \tag{2.17}$$

where $\vec{A} = \hat{\mathrm{D}}^2\vec{x}$ is the acceleration of O relative to $\hat{S}$.

This is useful because it can be used in principle to re-express the Kirchoff-Love equation (2.14) in a non-inertial frame local to the plate, as the acceleration arises naturally there. The acceleration formula (2.17) can then be used to to re-express Newton's second law in S rather than $\hat{S}$:

$$m\hat{\vec{a}} = \vec{F} \text{ in } \hat{S} \implies m\vec{a} = \vec{F} - \underbrace{m\,(\mathrm{D}\vec{\omega}) \times \vec{r}}_{\text{“Euler force”}} - \underbrace{2m\vec{\omega} \times \mathrm{D}\vec{r}}_{\text{“Coriolis force”}} - \underbrace{m\vec{\omega} \times (\vec{\omega} \times \vec{r})}_{\text{“Centrifugal force”}} + m\vec{A} \text{ in } S. \tag{2.18}$$

Since we are only concerned with elastic oscillations of the plate in the $\vec{e}_3$-direction, it suffices to take the dot product of equation (2.18) with $\vec{e}_3$ in order to obtain the equivalent of the Kirschoff-Love equation (2.14). We do not write down the equation in full, since it depends on the specific rotations being applied. However, we simply state that the term involving the second derivative of w_0 with respect to time corresponds to $\mathbf{a} \cdot \mathbf{e}_3$ and that the external loads $q(x, y, t)$ correspond to $\vec{F}$.

It is not quite enough to simply apply a change of frame to the Kirschoff-Love equation to take all physical considerations into account. There is also the weight of the plate, the inertia, the internal elastic forces inside the plate, and also the air drag. However, in a factory setting, the effects of these terms would be questionable, and may make the equations significantly more coupled without adding much insight. This is particularly the case for the air drag because a complete description would rely on coupling the elastic displacement equations discussed hitherto with the Navier-Stokes equations of fluid dynamics. A simplified starting point could be to introduce a simple drag law, where the drag force on an object in a fluid is proportional to the square of the speed at which object is being passed by the fluid. The exact nature of the physical effects is well beyond the scope of this report, but would be interesting nonetheless.

Another physical phenomenon to consider is that many elastic materials deform plastically when their displacements pass a limit that is large enough in size. This is often called *yield*, and a hypothesis governing an instance in which a material deforms is known as a *yield criterion*. Yield hypotheses are frequently given in terms of functions of the entries of the stress tensor σ; for example, the *von Mises' stress criterion* [6]

$$\sigma_{\mathrm{VM}} = \sqrt{\frac{3}{2}(\sigma_{11}^2 + 2\sigma_{12}^2 + \sigma_{22}^2) - \frac{1}{2}(\sigma_{11}^2 + \sigma_{22}^2)}.$$

This is a nonlinear equation on the stresses inside the elastic plate, and therefore the strains and the elastic displacements. The critical stress σ_{VM}, if exceeded by the elastic plate, will result in it failing to revert to its original state. The von Mises' stress criterion could be a useful diagnostic in designing possible paths for the robotic arm to take; since the stress tensor entries are determined from the elastic displacements, this could result in the suggestion of a suitable inverse problem: *given that the stress cannot exceed a certain limit, how should the path of the robotic arm be constructed?*

We also note that, in a factory setting, the time variation in the Kirschoff-Love equation (2.14) is likely to be relatively slow. This indicates that the time-dependence of q and the time-derivative can be neglected, although one needs to check that this is justified by nondimensionalising and using typical scalings provided by those at F.EE. Upon doing so, the Kirschoff-Love equation reduces to a biharmonic equation $D\nabla^4 w_0 = -q(x, y)$, which can be solved using asymptotic methods and possibly (depending on the form of q and the presence of the non-inertial terms) complex variable methods. A simplifying assumption could be to use the *quasi-static hypothesis*, which takes out the time dependence initially to find a solution, and then reinstates the time-dependence in q once this is done (for example, by seeking an asymptotic expansion solution to w_0 which is steady at leading-order).

2.3 The Numerical Simulation

In this section, we outline the steps towards a possible simulation with a simple model of a solid shell using Finite Element Analysis, FEA (See [7], for the application of FEA to solid mechanics).

The purposed model consists of a shell of small thickness moving on its transversal direction (see Fig. 2.4), z-axis in the figure, with constant acceleration, $\vec{a} = (0, 0, a_z)$. We consider non-inertial forces, the weight of the plate, and friction with the air.

This model is suitable to be analyzed by means of the FEA, for which the main steps are: meshing the domain, then introduce the weak formulation to set up the local equations for the elements, that are coupled to form the *global system*. The boundary conditions of the original problem are imposed on this system of linear equations to give a *reduced system* whose solution, the *nodal solution*, consists of the displacements of the nodes with respect to the undeformed body. (and other quantities, depending on the element type chosen to carry out the analysis, see below). Later, in the post-process, one can compute other results derived from the nodal solution. Typically, the strains and stresses and the Von Mises' stress.

Weak Formulation and the Principle of Virtual Work

The formulation of the weak form of the equations from the principle of virtual work is presented in many books in solid mechanics. Here, we give a brief account and point the reader to [8, Chap. 1], and references therein.

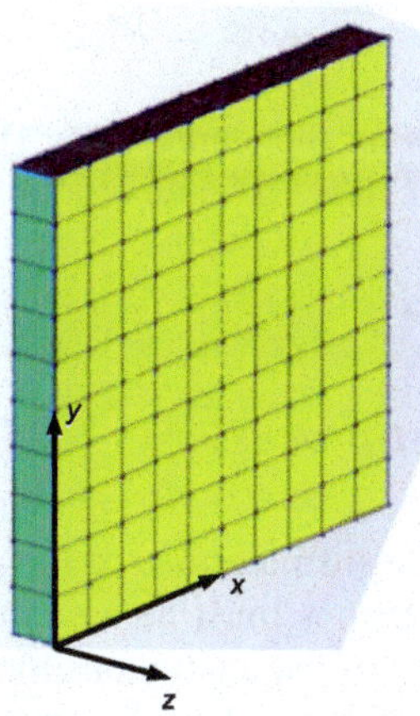

Fig. 2.4 Meshing the 3D domain: element and nodes. *Source* https://openfoamwiki.net

So, basically, to find the local equations of the element, $\mathcal{P}^e$, the underlying idea is that the changes in its internal strain energy are equal to the virtual work of the external forces acting on it. Therefore,

$$\underbrace{\iiint_{\mathcal{P}^e} \delta\varepsilon^\top \sigma \, \mathrm{d}V}_{\substack{\text{Virtual changes in}\\ \text{the internal strain}\\ \text{energy}}} = \underbrace{\iiint_{\mathcal{P}^e} \delta\ell^\top f_b \, \mathrm{d}V + \iint_{\partial\mathcal{P}^e} \delta\ell^\top f_t \, \mathrm{d}S + \ldots}_{\substack{\text{Virtual work made by}\\ \text{the external forces}}}, \tag{2.19}$$

with ε being the strain, σ the stress, ℓ the vector displacement of the nodes, f_b the body forces (weight and non-inertial forces) and f_t the surface forces (note that the integral is taken on the boundary of the element) whereas $\delta\varepsilon$ and $\delta\ell$ denote the virtual changes on strain and displacements, respectively.

In a linear analysis context, the strain and displacements are related by the strain-displacement matrix, B, as $\varepsilon = B\ell$, so for virtual displacements, $\delta\varepsilon = B\delta\ell$. On the other hand, the stress-strain law states that $\sigma = C\varepsilon$. Then, the substitution of these relations in (2.19) yields

$$\delta\ell^\top \underbrace{\left(\iiint_{\mathcal{P}^e} B^\top C B \, \mathrm{d}V\right)}_{\substack{\|\\ K^e}} \ell^e = \delta\ell^\top \underbrace{\left(\iiint_{\mathcal{P}^e} f_b \, \mathrm{d}V + \iint_{\partial\mathcal{P}^e} f_t \, \mathrm{d}S + \ldots\right)}_{\substack{\|\\ R^e_{ext}}},$$

and (arbitrary) virtual displacements, $\delta\ell$, cancel out to give,

$$K^e \ell^e = R^e_{ext}, \tag{2.20}$$

which are the local equations for the element $\mathcal{P}^e$. The integrals are computed using Gaussian integration (see [9, Chap. 5], for an account of Gaussian integration in FEA related problems).

Nodal Solution and Post-process

Note that K^e and R^e_{ext} are the stiffness matrix and the force vector of element $\mathcal{P}^e$, respectively. These local equations must be assembled to form the global system. Next, the boundary conditions of the continuous model must be tailored to be set at the (global) nodes of the assembled system. No traction nor compression is assumed to act at the boundaries, so we consider only *essential* boundary conditions: the displacement of the nodes grabbed by the grips are fixed to zero. The outcome is a *reduced* system of linear equations, where only the displacements of *free* (not fixed) nodes remain as unknowns. In the end, at the post-process, the strain and the stress at the nodes are computed from the displacements using the above-mentioned relations, i.e.,

$$\varepsilon^e = B\ell^e, \qquad \sigma^e = C\varepsilon^e = CB\ell^e;$$

moreover, von Mises' stress σ^e_{VM} can be derived from the components of σ^e.

Meshing the 3D Domain and MITC9 Shell Element

In FEA, it is worth choosing the most suitable element type to mesh the domain. In the case of a car's bodywork, where the thickness is small in front of the surface extension, and curves surfaces are present, shell finite elements are widely used. For a description, see [9, Chap. 5]; and [4, Chap. 11], for a thorough analysis of shell elements and a complete list of references. In [8, Chap. 3], IDC (Isoparametric Degenerate Continuum) and MITC9 (Mixed Interpolation of Tensorial components with 9 nodes) shell elements are explained in more detail. In particular, we propose the use of this last one to analyze the car's bodywork. The formulation of the MITC9 finite element we comment below is taken from a more recent work [10], where this element is extended to the elastoplastic analysis of shell structures by adding and hysteresis model.

We remark that, as pointed in [8], the main characteristic of the MITC family of finite elements is that, in order to avoid the shear-locking phenomenon (due to the small thickness, nodes in that direction are very close and the associated coefficients in the stiffness matrix become very large), it uses separate interpolation functions for the tangent and transverse shear strain components.

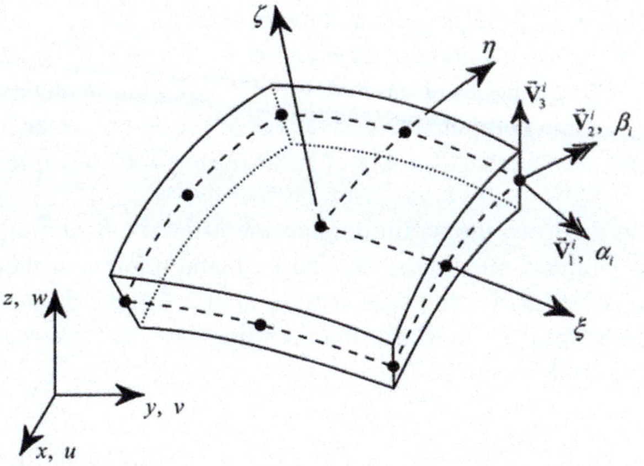

Fig. 2.5 MITC9 element. *Source* A. N. Moysidis and V. K. Koumousis (see [10])

The aspect of the MITC9 element is shown in Fig. 2.5. The 9 nodes are placed in the middle surface. The coordinates ξ, η, and ζ are local coordinates "attached" to the central node; ξ, η are curvilinear coordinates whereas axis ζ points in the direction of the thickness (therefore, it is not exactly normal to the mid-surface). Furthermore, associated with each node, there are three orthogonal vectors V_{1i}, V_{2i} and V_{3i}. For each $i = 1, \ldots, 9$, V_{3i} is the vector on the line defined by the thickness from the bottom to the top of the element. If $p_i^\top = (x_i, y_i, z_i)$, $i = 1, \ldots, 9$, are the Cartesian coordinates of the nodes, then the Cartesian coordinates of any point of the element are given in terms of the local element coordinates ξ, η, ζ, by

$$\begin{pmatrix} x \\ y \\ z \end{pmatrix} = \sum_{i=1}^{9} N_i(\xi, \eta) \begin{pmatrix} x_i \\ y_i \\ z_i \end{pmatrix} + \frac{\zeta}{2} \sum_{i=1}^{9} N_i(\xi, \eta) V_{3i}, \tag{2.21}$$

here $N_i(\xi, \eta)$, $i = 1, \ldots, 9$, are the shape functions (see [8, Chap. 3]). To define the two vectors orthogonal to each V_{3i} one way to proceed is to take $V_{i1} = \hat{j} \times V_{3i}$, $V_{1i} = V_{3i} \times V_{1i}$, where $\hat{j}$ is the unit vector in the direction of the y-axis. Let $\hat{v}_{1i}$, $\hat{v}_{2i}$ be the normalized vectors of V_{1i} and V_{2i}, and α_i, β_i be rotations of vector V_{3i} about $\hat{v}_{1i}$ and $\hat{v}_{2i}$, respectively. If in addition, it is assumed that the angles α_i, β_i, $i = 1, \ldots, 9$ and the strains in the direction normal to the middle surface are small, then the displacements u, v, w in the Cartesian coordinates x, y, z (see Fig. 2.5) are given by the formula,

$$\begin{pmatrix} u \\ v \\ w \end{pmatrix} = \sum_{i=1}^{9} N_i(\xi, \eta) \begin{pmatrix} u_i \\ v_i \\ w_i \end{pmatrix} + \frac{\zeta}{2} \sum_{i=1}^{9} t_i N_i(\xi, \eta) \left(-\alpha_i \hat{v}_{2i} + \beta_i \hat{v}_{1i}\right), \tag{2.22}$$

here t_i is the thickness of the element at the i-th node for $i = 1, \ldots, 9$. Thus, the displacements (u, v, w) for any point of the element are determined by the displacements $(u_i, v_i, w_i)_{i=1,\ldots,9}$ of the nodes at the mid-surface and rotation angles $(\alpha_i, \beta_i)_{i=1,\ldots,9}$.

From (2.22) we can compute the derivatives of the displacements with respect to the local element coordinates: $\partial_\xi u(\xi, \eta, \zeta)$, $\partial_\eta u(\xi, \eta, \zeta)$, $\partial_\zeta u(\xi, \eta, \zeta), \ldots$ (the same for v and w).

Finally, to write down the strain displacement-matrix B and later compute the stiffness matrix of the element (see [8, Chap. 3], and references therein for explicit formulas of B as well as for the stress-strain matrix C), the derivatives of the displacements with respect to the Cartesian coordinates x, y, z are required, but these follow at once from the relations

$$\nabla \hat{u}(x, y, z) = \left[\frac{\partial(x, y, z)}{\partial(\xi, \eta, \zeta)}\right]^{-\top} \nabla u(\xi, \eta, \zeta), \ldots \text{ (the same for } \hat{v} \text{ and } \hat{w}),$$

where $\hat{u}, \hat{v}, \hat{w}$ are the displacements as functions of the Cartesian coordinates, ∇ is the usual notation for the gradient operator (actually, $\nabla = (\partial/\partial x, \partial/\partial y, \partial/\partial z)^\top$ at the l.h.s. and $\nabla = (\partial/\partial \xi, \partial/\partial \eta, \partial/\partial \zeta)^\top$ at the r.h.s), and $[\partial(x, y, z)/\partial(\xi, \eta, \zeta)]^{-\top}$ is the inverse and transposed Jacobian matrix of the transformation (2.21).

Remarks on Software. Although general-purpose programming languages, such as FORTRAN and MATLAB have been successfully applied to FEA (see, for example, [9, Chap. 12], [11]) alternatively, one can use specialized software like ANSYS (see [12]) or ADINA (which has the MITC9 shell element available out-of-the-box, see [13]), or, try software from other projects that have more recently come out, such as FEniCS (see [14]), or, Firedrake (see https://firedrakeproject.org/).

2.4 A Simple Model Problem

Here we present a simple model problem to aid analysis. In order to do this we will consider a 1D analog to the problem presented in the earlier sections. We will consider a thin rod, clamped at one end, that is free to move and rotate in a single plane (see Fig. 2.6).

The appropriate model to describe the deformation of the rod is the Euler-Bernoulli equation. As with the previous model for a 2D plate we will orient our axis with respect to the rod; with the origin being the clamped end of the rod; x being the direction parallel to the undisplaced rod; z being the direction perpendicular to the undisplaced rod, in the plane of motion of r and θ; and y is the direction perpendicular to the rod, into the page in Fig. 2.6. The Euler-Bernoulli equation reads

$$EI \frac{\partial^4 w}{\partial x^4} = q - \rho \frac{\partial^2 w}{\partial t^2},$$

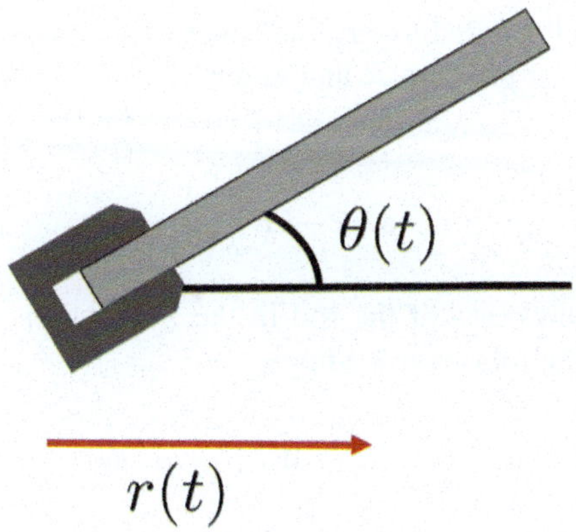

Fig. 2.6 Schematic of model problem, showing rod, light grey, clamped at one end

where E is the elastic modulus, $I = \iint z^2 dydz$, is the second moment of area, w is the displacement of the rod in the direction perpendicular to the rod, q is the load on the rod (the force applied to the rod in the perpendicular direction), ρ is the density per unit length of the rod, and t is time. EI is often referred to as the flexural rigidity, and is the 1D analog to the bending stiffness, D, in the Kirchoff-Love equation. We must establish the form of q due to the motion of the rod. We will omit the derivation here, as it is functionally the same as the derivation in the 2D plate case. The form that the load takes is

$$q = \rho\left[\ddot{r}\sin(\theta) - g\cos(\theta) + \dot{\theta}^2 w - \ddot{\theta}x\right],$$

where the terms are, moving left to right, the acceleration force, the force due to gravity, the Centrifugal force, and the Euler force. The appropriate boundary conditions for w are

$$\begin{aligned} w(x,0) &= \frac{\partial w}{\partial t}(x,0) = 0, \text{ initially undeformed and stationary,} \\ w(0,t) &= \frac{\partial w}{\partial x}(0,t) = 0, \text{ clamped at one end,} \\ \frac{\partial^2 w}{\partial x^2}(L,t) &= \frac{\partial^3 w}{\partial x^3}(L,t) = 0, \text{ free at one end,} \end{aligned}$$

where L is the length of the rod. Whilst we are currently imposing r and θ, finding the stress for a given path, a piece of further work will be to find the path that minimises the stress. With this in mind we will give boundary conditions for θ and r

$$\begin{aligned} &\theta(0) = \theta_0, \quad \dot{\theta}(0) = 0, \quad \theta(T) = \theta_1, \quad \dot{\theta}(T) = 0, \\ &r(0) = 0, \quad \dot{r}(0) = 0, \quad r(T) = R, \quad \dot{r}(T) = 0. \end{aligned}$$

In other words, the rod is initially at rest at some specific angle, θ_1, and ends at rest, at some later time, T, at position R and angle θ_1. Finally we have the form of the stress, σ,

$$\sigma = -hE\frac{\partial^2 w}{\partial x^2},$$

where h is half the thickness of the rod in the z direction. We now seek to non-dimensionalise with the following scalings

$$x = L\hat{x}, \quad t = T\hat{t}, \quad w = W\hat{w}, \quad r = R\hat{r}, \quad \sigma = \frac{hE}{L}\hat{\sigma},$$

where W is a characteristic lengthscale for the displacement. This gives us the following, dropping the hats for convenience,

$$\lambda\frac{\partial^4 w}{\partial x^4} = \mu\ddot{r}\sin(\theta) - \frac{1}{\mathrm{Fr}^2}\cos(\theta) + \dot{\theta}^2 w - \nu\ddot{\theta}x - \frac{\partial^2 w}{\partial t^2},$$
$$\sigma = -\frac{\partial^2 w}{\partial x^2},$$

where $\lambda = \frac{EIT^2}{\mu L^4}$, $\mathrm{Fr} = \sqrt{W/gT^2}$, $\mu = R/W$, $\nu = L/W$ are our non-dimensional groups, together with the initial and boundary conditions

$$w|_{x=0} = 0, \quad \left.\frac{\partial w}{\partial x}\right|_{x=0} = 0, \quad \left.\frac{\partial^2 w}{\partial x^2}\right|_{x=1} = 0, \quad \left.\frac{\partial^3 w}{\partial x^3}\right|_{x=1} = 0,$$
$$w|_{t=0} = 0, \quad \left.\frac{\partial w}{\partial t}\right|_{t=0} = 0,$$
$$\theta(0) = \theta_0, \quad \dot{\theta}(0) = 0, \quad \theta(1) = \theta_1, \quad \dot{\theta}(1) = 0,$$
$$r(0) = 0, \quad \dot{r}(0) = 0, \quad r(1) = 1, \quad \dot{r}(1) = 0.$$

Now we will now simplify this further by assuming w does not depend on t, Fr is large, and $\ddot{\theta} = \ddot{r} = 0$. This gives us

$$\lambda\frac{\partial^4 w}{\partial x^4} = \dot{\theta}^2 w,$$
$$w = \frac{\partial w}{\partial x} = 0 \text{ at } x = 0,$$
$$\frac{\partial^2 w}{\partial x^2} = \frac{\partial^3 w}{\partial x^3} = 0 \text{ at } x = 1.$$

We can define $\beta = \left(\frac{\dot{\theta}^2}{\lambda}\right)^{1/4}$. The above problem only has non-trivial solutions if

$$\cos(\beta)\cosh(\beta) + 1 = 0,$$

in which case the solution is

$$w = A\left[\cosh(\beta x) - \cos(\beta x) + \frac{\cosh(\beta) + \cos(\beta_n)}{\sin(\beta) + \sinh(\beta)}\left(\sin(\beta x) - \sinh(\beta x)\right)\right].$$

It is worth noting that the non-trivial solutions occur precisely when the beam is rotating at one of its natural frequencies, the resulting solution is an example of resonance.

2.5 Conclusion and Further Work

To continue with the work we have started, we understand that both the numerical implementation of the finite element approach that we proposed in Sect. 2.3 and a deeper analysis of the simple model considered in Sect. 2.4 have to be done. The finite element analysis can also be done by using some commercial software, like ANSYS (https://ansys.com), with a comparison of results. It is clear that the commercial software will have some advantages in the beginning, but the implementation of *ad hoc* models, for example, for the air resistance over the pieces, will perhaps be more difficult. A long list of alternatives to ANSYS can be found in Wikipedia article https://en.wikipedia.org/wiki/List_of_computer_simulation_software.

This last-mentioned is one of the major issues that remain unsolved, i.e., how to model the air resistance. One possibility would be to couple the solid mechanics calculations with fluid mechanics calculations; doing so would make the problem more complete, but at the same time very complex. Furthermore, reasonable and simple *ad hoc* models, perhaps based on Asymptotic Analysis techniques, should exist or could be designed. In our opinion, performing numerical experiments as well as the laboratory experiments in this direction should help in addressing these challenges.

The analysis of the simple problem stated in Sect. 2.4 would also be important. It would, for example, simplify the analysis of the role of the so-called *jerk*, or, the first-derivative of the acceleration, on the deformations. It would also be useful to understand the differences between the quasi-static case (time-dependent exterior forces, but no inertial forces) and the true-dynamic case (including the inertia). It is clear that in fast movements of heavy bodies, the inertial forces will be very relevant, but the important thing here is to know the limits of this reasonable simplification.

This analysis of the *jerk* would also help in the direction of the design of the geometries of trajectories, to avoid non-progressive changes in accelerations. The changes in the accelerations are precisely the *jerk*. This is a classical problem in railway design, and relation to their results should be discussed.

Acknowledgements T.B. & O.B. acknowledge the support provided by the EPSRC Centre for Doctoral Training in Industrially Focused Mathematical Modelling (EP/L015803/1). J.R.P. is supported by MINECO (Spain) grant MTM PGC2018-100928-B-I00. S.K. is supported by the grant Severo Ochoa SEV-2017-0718. J.S.-M. acknowledges partial support by MINECO (Spain) grant MTM2017-84214-C2-1-P.

References

1. D.E. Whitney, The mathematics of coordinated control of prosthetic arms and manipulators (1972)
2. V. Braibant, M. Geradin, Optimum path planning of robot arms. Robotica **5**(4), 323–331 (1987)
3. P. Howell, G. Kozyreff, J. Ockendon, *Applied Solid Mechanics*, vol. 43 (Cambridge University Press, 2009)
4. J. Reddy, *Theory and Analysis of Elastic Plates and Shells*, 2nd ed. (CRC Press, 2007)
5. H. Goldstein, C.P. Poole, J.L. Safko, *Classical Mechanics* 3rd edn. Chapter 7 (2002)
6. R.V. Mises, Mechanik der festen körper im plastisch-deformablen zustand. Nachrichten von der Gesellschaft der Wissenschaften zu Göttingen, Mathematisch-Physikalische Klasse, vol. 4 (1913), pp. 582–592
7. O.C. Zienkiewicz, R.L. Taylor, *The Finite Element Method*, vol. 2, 5th ed. (Butterworth Heinemann, Oxford, 2000)
8. A.M. Lush, Coupling of beam and shell finite elements for the rapid analysis of tubular structures, MSc Thesis, School of Engineering and Computing Sciences. Durham University (2014)
9. K. Bathe, *Finite Element Procedures* (Prentice Hall, 2006)
10. A. Moysidis, V. Koumousis, Hysteretic shell finite element. J. Eng. Mech. **145**(5) (2019)
11. Y.W. Kwon, H. Bang, *The Finite Element Method using MATLAB*. Series CRC Mechanical Engineering Series, 2nd ed. (Chapman & Hall/CRC, Boca Raton, 2000)
12. Wikipedia Contributors, Ansys — Wikipedia, the free encyclopedia, https://en.wikipedia.org/w/index.php?title=Ansys&oldid=963235585, 2020, [Online; Accessed 23-June-2020]. [Online; Accessed 23-June-2020]
13. Wikipedia Contributors, ADINA — Wikipedia, the free encyclopedia (2018), https://en.wikipedia.org/w/index.php?title=ADINA&oldid=875541407
14. Wikipedia Contributors, FEniCS Project — Wikipedia, the free encyclopedia (2020), https://en.wikipedia.org/w/index.php?title=FEniCS_Project&oldid=957113708 [Online; accessed 23-June-2020]

Chapter 3
Early Estimation of Level Difficulty in Mobile-Games

Maria Aguareles, Llorenç Badiella, Jordi Baró, Mar Jovani, Daniel Lopez, Alejandro Molero, James Morrill, Mohammad Peyman, Aleix Ruiz de Villa, and Rahil Sachak-Pratwa

Statement of the Problem

At King, levels are at the core of our games. Finding the right balance in difficulty, though, is quite a challenge! The problem we present here aims at helping our Level Designers estimate how hard a level will be before it reaches our players. We will provide a dataset with different level definitions from a real game, descriptions of how the ingredients interact with each other, and a target indicator that should be closely related to the level difficulty. Now, is it possible for us to estimate this indicator given a brand new level?

For the problem we have at hand, we will be focusing on a specific game: *Bubble Witch 3 Saga*. BW3S was developed in Barcelona and released worldwide in 2017. It has a stable player base, with millions of daily active users.

The game consists in a linear progression map with levels that the player needs to beat in order to move forward. Each level represents a puzzle that has to be completed with a limited number of moves. See Fig. 3.1, which corresponds to the first puzzle of the game. The board shows a configuration of colored bubbles and, in this case,

M. Aguareles (✉)
Department d'Informatica, Matematica Aplicada i Estadistica, Universitat de Girona, Girona, Spain
e-mail: maria.aguareles@udg.edu

L. Badiella · M. Peyman · A. R. de Villa
Servei d'Estadística Aplicada de UAB, Barcelona, Spain

L. Badiella
Universitat Autònoma de Barcelona, Bellaterra, Spain

J. Baró · D. Lopez · A. Molero
Centre de Recerca Matemàtica, Campus de Bellaterra, Barcelona, Spain

M. Jovani · R. Sachak-Pratwa
Universitat Politècnica de Barcelona, Barcelona, Spain

J. Morrill
University of Oxford, Oxford, England

M. Aguareles et al. (eds.), *Applications of Industrial Mathematics*,
RSME Springer Series 8, https://doi.org/10.1007/978-3-031-32130-6_3

Fig. 3.1 BW3S level 1

some of them have owls trapped inside. The goal of this first puzzle is to "free the owls".

To do so, our main character Stella will be shooting bubbles that, if matched with the same color on the board, will pop all the connected bubbles of that color and release the owls inside. In the picture, the represented shot will send a yellow bubble towards a yellow configuration that will break the first triangular section and release one owl.

In the shooting area below, we also see that we have 37 remaining bubbles to shoot (usually referred to as *moves*) and that the next bubble that we will be shooting is blue. As one could expect, the more moves you have, the easier the puzzle is.

If we can complete the mission (in this case, by freeing all 6 owls) with the moves that we have available, we will move onto level 2. Otherwise, we will need to try again until we succeed.

Puzzles can be of different types:

- **Collect (free the owls)**: As in Level 1 above, the goal is to pop all the bubbles that have an owl trapped inside. You need to release all the owls to pass the level.
- **Clear All**: The goal here is just to pop all the bubbles on the board. The first puzzle of this type that we will find is in Level 4.
- **Boss**: These puzzles will require players to defeat Wilbur, Stella's enemy, by popping bubbles that have *fairies* inside. These fairies will "attack" Wilbur and drain his life bar until it reaches zero. The first puzzle of this type is Level 10.

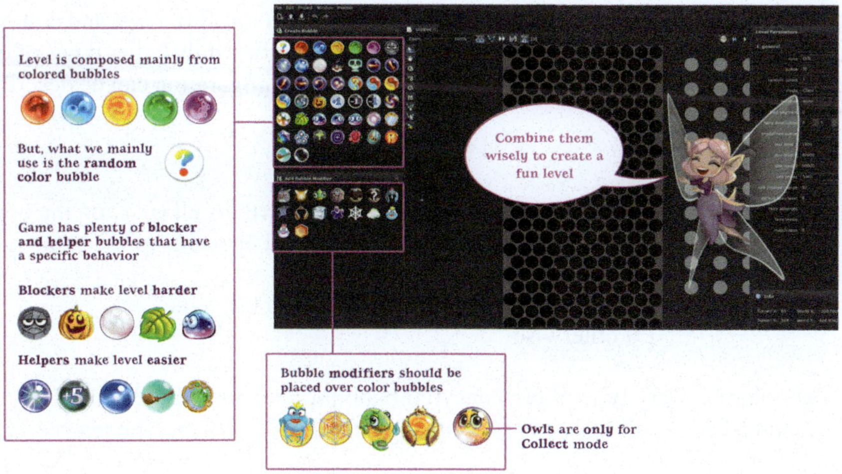

Fig. 3.2 Level editor

- **Ghost**: In this last puzzle type we will find a floating ghost that will try to go up in the board. Our goal is to clear its path by popping bubbles. Once the ghost can reach the top of the board, the level is complete. The first puzzle of this type can be found in Level 11.

In this project, for the sake of simplicity, we will only focus on the **Clear All** puzzle type.

Level Layout As you progress through the map, the level configuration can get more intricate. Level Designers use an editor tool that help them build new puzzles. Figure 3.2 shows what the editor looks like. They have an empty board and a list of elements that they can place in each cell.

These elements can be classified in four different groups:

- **Bubbles**: The main ingredient. Bubbles can be of a fixed color or random. In the latter case, every time that a player starts a level, the color of the bubble can be different (among the colors that are used on the board).
- **Blockers**: Cells can have different types of blockers, which basically complicate things for the player. They can be, for instance, obstacles that need to be cleared before you can shoot through them.
- **Helpers**: These cells will help players towards their objective. For instance, they can create explosions that pop additional bubbles, increment the number of moves available, etc.
- **Modifiers**: Modifiers can be placed over colored bubbles to change their behavior in a certain way. For instance, in Collect-type puzzles, owls inside a colored bubble are a modifier.

Definition of the Level Let $n, m > 0$ be the maximum number of rows and columns respectively that a board can have. Let $l > 0$ be the number of different types of elements that we can put into a given cell. Then, a board configuration can be described by

$$A = (a_{i,j}^k)_{i,j,k},$$

where $i \in \{1, .., n\}$ and $j \in \{1, .., m\}$ correspond to a certain cell (i, j) on the board and $k \in \{1, .., l\}$ to a certain "layer" that will contain all elements of the same type. Then,

$$a_{i,j}^k = \begin{cases} 1 \text{ if there is an element of type } k \text{ in the cell } (i, j), \\ 0 \text{ otherwise.} \end{cases}$$

In other words, every type of element (red bubbles, blue bubbles, random bubbles, owls, pumpkins...) will have an associated layer $k \in \{1, ..., l\}$, and the matrix

$$A^k = (a_{i,j}^k)_{i,j} = \begin{pmatrix} a_{1,1}^k & \cdots & a_{1,m}^k \\ \vdots & \ddots & \vdots \\ a_{n,1}^k & \cdots & a_{n,m}^k \end{pmatrix}$$

will encode with 1's their positions on the board. If the level doesn't have any element of type k, then A^k will be a zero-matrix.

If we go back to our first example in Fig. 3.1, where the level layout is quite simple, we would only have 5 non-zero layers[1] $A^{k_1}, ..., A^{k_5}$: one layer will have the position of the blue bubbles, another for the red ones, another for the yellow ones, one for the fairies and the last one for the owls.

Note that, given a cell (i, j), the sum

$$o(i, j) = \sum_{k=1}^{l} a_{i,j}^k$$

denotes the number of overlapping elements in that cell. In Fig. 3.1, wherever we only have a colored bubble, $o(i, j) = 1$, but for bubbles with an owl inside, the value is 2. Having $o(i, j) = 0$ means that the cell (i, j) is empty. Some layers exclude each other (for instance, there cannot be a red and a blue bubble in the same cell) but some combinations with modifiers, helpers or blockers are possible.

Figure 3.3 corresponds to Level 52. Here you can see that we have cells with

- Colored bubbles only, probably randomly colored in this case.
- Colored bubbles with a bat on top (an overlapping modifier that changes the behavior of the bubble when popped).
- Colored bubbles hidden behind a cloud (another modifier that can be used to hide the underlying element).

[1] In Level 1, all bubbles are of a fixed color, as it is a tutorial level with a fixed structure. If we also had some randomly colored bubbles, they would be part of a different layer.

Fig. 3.3 BW3S level 52

- Colored bubbles with fairies inside. Fairies are helpers that can be placed inside bubbles and, when popped, fly around and help you destroy other elements of the board.
- Colored bubbles with both fairies inside and a bat on top. These cells would have $o(i, j) = 3$.

Problem Statement We will provide a set of 130 real level configurations, all of the puzzle type Clear All, together with the rest of the parameters that define a level (number of playable colors –that will determine the randomly colored bubbles–, number of moves, etc...). We will also describe the dynamics and effects that each game element has. Finally, we will provide the *average pass rate* that we have observed in those levels.

The goal is to find a way to estimate some kind of "difficulty indicator", that should be correlated with the pass rate, and that can be computed given the definition of a new level.

3.1 Introduction

Digital entertainment (DE) constitutes a dynamic industry sector generating annually 2 trillion USD in revenue worldwide [8]. Among the broad offer of services, DE products compete for the time of the consumer, a.k.a the player in the context of casual DE products. Since DE do not cover primary needs, the main problem addressed by creators is to generate value to engage the consumer. A fundamental hypothesis in the Mechanics-Dynamics-Aesthetics (MDA) framework of game design [4] states that players are engaged in interactive DE media due to subjective intrinsic rewards emerging from a series of values, i.e. the aesthetics of play, covering secondary psychological and sociological needs.

One of the core aesthetics of play is an adequate equilibrium between challenge and skill [5]. This "Challenge" aesthetic is linked to the concept of flow within the framework of behavioural psychology [2]. We experience a sense of satisfaction at the optimal equilibrium between our skills being tested and the difficulty of the task, leading to a state of hyperfocus or flow, where our mind reduces the influence of external stimuli unrelated to the task at hand. Flow exists within a fragile equilibrium between challenge and skill. If the task is perceived as too difficult it generates a sense of frustration. If it is perceived as too easy it generates a sense of boredom and banality. This state of hyperfocus, i.e. flow, emerges as a consequence of complex neurocognitive processes related to natural learning mechanisms in primates [3] and the experience appears to be susceptible of a physiological predisposition [1].

As a consequence, as all aesthetic values, Challenge is intrinsically subjective due to the diversity of predispositions and skills among players. The player response to an interactive DE product in terms of challenge experience is difficult to forecast. In particular, the level of Challenge, or difficulty, and skill are hard to quantify by using straightforward measures independently. However, big data sets recorded during play-through can help us understand the challenge perception from the player decisions during the game, i.e. the dynamics in terms of MDA, and progress.

In the current report we explore several ways to produce an early estimation of the level of difficulty in mobile-games. For the study we take data sets of the game Bubble Witch 3 Saga from the company KING. In this case, a player has to win a sequence of levels. All players have the same levels. A level is totally characterised by a set of features. The levels can be arbitrarily labelled as 1, 2 etc in order to describe the unlocking order. A player has to win level 1, in order to unlock level 2. A player can try to win an unconstrained number of times. When a player wins a level they can then access the following level. When a player tries to win a level, we say it is a trial. The challenge is to discover patterns to relate the characteristics of a level and the difficulty of the level. The characteristics of the level are described as a high-dimensional dataset and the difficulty is described by means of the number of trials.

Given a level and a player, for each trial, the player has to clean the screen with a number of shots smaller than a fixed number, N_0. Remark, N_0 depends on the level, but it does not depend on either the player or the trial.

3.2 Mathematical Characterisation of the Challenge

We work under the assumption of the existence of a functional relation, f, between the design and the difficulty of a level, i.e.,

- $X =$ Design of the level ($design$)
- $Y =$ Difficulty of the level ($difficulty$)

so $Y = f(X)$. The challenge is to search for a pattern to describe f through a sample set of levels where the complete design and the difficulty for each one is reported. A dataset is the main tool to develop the solution, then several steps have to be considered in order to check the capabilities of the data to solve the problem.

The data characterising the design of the level is classified in three subsets.

- Level identifiers: $levelid$, which is just a number indicating the level.
- Level Parameters: $queue$, which encodes the particular ordered series of balls that sequentially appear, and $covariates$, which contains other extra information on the sequence of balls which may have an effect on level difficulty, like for instance the number of ball modifiers in a given ball queue.
- Level Graph Elements: $configuration$, which encodes the particular initial puzzle to be completed.

The variable $difficulty$ contained in Y is the target variable we want to predict through the "pass time". It represents how difficult it is to beat the level expressed as an average number of attempts. A pass time of 5 means players spend on average 5 attempts to win.

Each subset reports a different type of data and the role in the searched pattern can be different. The main interest is to know the effect of the design on the difficulty. If $configuration$ is fixed then there is a monotonic relationship between $difficulty$ and every one of the other variables ($levelid$, $queue$ and $covariates$). The sets $covariates$ and $configuration$ share common information. The information given by pass-time includes the relationship between the experience of a player and the level difficulty, but these effects cannot be distinguished given the data available. Finally, for a fixed $configuration$ and a fixed $levelid$ (or experience), the $configuration$ (and/or $covariates$) produces a description of the $difficulty$ as a function of $queue$.

The dataset for describing the screen configuration contains more than one thousand variables, and only one hundred cases are reported. Thus, a reduction of the dataset has to be considered.

On the Difficulty of the Level

The difficulty of a level is defined by the average number of trials that a player needs in order to win. More precisely, for a fixed level and fixed $queue$, if

$$T = \text{number of trials needed to win},$$

win is the event $N < queue$ and

$$N = \text{number of needed shots to pass the level (without } queue\text{)},$$

the difficulty, Y is provided by the estimation of the expected value, $E(T)$, through the sample mean, $\bar{T}$, so that Y ends up being a function of $\bar{T}$.

3.3 Modelling Strategy

The next steps summarise the main approach developed in the study group.

- Step 1. Consider $X = configuration + queue$ since the effect of X on Y can then be factorised.
- Step 2. Reduce the dimension of $configuration$ to produce a new set of $newFeatures$.
- Step 3. Develop a deterministic relationship between Y and $queue$, through probability modelling. The probability model can introduce a new set of parameters, θ. That is,

$$Y = g(queue, \theta),$$

  where g is a deterministic function.
- Step 4. Model θ with $newFeatures$, $covariates$ and $queue$.

The aim of this approach is to reduce two problems of the current dataset, i.e. the representativeness and the high-dimensional features effects of the dataset. The more complex the probabilistic model, the more these problems will be reduced.

In order to show the steps of this strategy, an easy approach for the probability model was considered. The exponential families are the first option to search a parametric probability model [6].

We start by transforming the response and we consider the independence between winning in different games. Then if $T \sim \text{Geom}(p)$ for a fixed probability p, then $Y = 1/p$.

Then, data was explored in order to identify the approach to reduce the dimension of X. Some assumptions were stated:

- According to the first exploration, the number of blockers and helpers is important.
- Distribution of elements in the board might be very relevant.
- The relationship between elements might play an important role as well (for instance, how close to one another are different elements on the board).

The reduction process is based on graph construction (embedded in the screen) where nodes recover the cluster structure of the bubbles and the edges recover the adherence to these. To summarise the information of the graph, it was simplified to 9 regions. At the end 72 features were obtained. In each region, we will look at 8 features: Number of colour clusters, Number of blocker clusters, Number of exploding helper clusters,

Number of non-exploding helper clusters, Density of fairies, Density of non-fairy modifiers, Connections of blockers and Connections of exploding helpers.

The strength of this approach for solving the challenge is based on the recovery of the probability model. Additional advantages are produced, since additional information will be obtained from the results. For instance, improving queue sampling using Bayesian methods.

3.4 Data Analysis and Learning Models

Description of the Dataset

Bubble Witch 3 Saga has, to this date, more than 2000 levels, divided in more than 100 episodes. Each level can be of a different type, where each type requires a specific goal to be fulfilled in order to complete the level. The levels we have worked with are all of the *Clear All* type, where the goal is to erase all the color bubbles that appear on the screen. Thus, we would like to remark that all the analysis and considerations we have done are specific for that level type. In other words, trying to estimate the difficulty of levels of a different type would require additional considerations, or possibly a completely different approach.

Our dataset consists of a collection of properties that characterize the first 130 *Clear All* levels of the game, together with the observed difficulty of that level, expressed as an average number of attempts required in order to complete it. In principle, the higher the average number of attempts required, the more difficult the level would be. This dataset is initially organized as a table with 130 rows (one for each level) and 38 columns (one for each property, plus the average number of attempts). Some of these columns contain numerical variables that define features such as the number of available bubbles that the player can shoot in order to complete the level, or the amount of different colors that the bubbles can have. But the majority of these columns (29 of them) contain, in fact, matrices that describe the position of different elements in the board, where each column corresponds to one specific element type. In general terms, each element type falls into one of these categories:

- **Color bubbles**, which are the main elements that appear in the screen. They can be either red, blue, yellow, green or purple. There are also random color bubbles, which are bubbles whose color is redefined every time the level is loaded.
- **Blockers**, which are elements that add difficulty to the game by, for instance, preventing the player from shooting to a particular position, adding new bubbles, or even decreasing the number of available shoots.
- **Helpers**, which can make the game easier. Examples would be elements that explode when hit, destroying bubbles around them, or that change the behavior of bubbles or blockers, among others.
- **Modifiers**, which add extra properties to existing bubbles. Some modifiers are helpful, while some others make the game more complicated.

The entries of these matrices are either 1's or 0's, where a 1 in a particular position would indicate that an element of a certain type is present on that position on the board. Since the board has 85 rows and 11 columns, each matrix can be represented as a 935-dimensional vector. Therefore, we have, in fact, a total amount of 27115 features associated to each level. With only 130 data points, using all these features directly in order to fit any model would inevitably lead to over-parametrization.

Therefore, the main problem we have tried to solve is the definition of a reduced set of features, small enough to be used as a way to train a model, but large enough to correctly represent the characteristics of each level so that it can be used to make correct predictions of its difficulty. By "small enough", we could think that any number of features below 100 could be helpful enough (given the number of levels available), but we have preferred to try to significantly reduce that number as much as possible. We had two reasons for that. First of all, it is good practice to start with models that are as simple as possible. Secondly, from a data scientist point of view, one should not only try to use data to obtain results, but also to extract knowledge from it that can be useful later on. In that sense, having a small set of features helps to gain insight of the underlying factors that might affect difficulty, specially if those features are easy to understand.

Graphs of Clusters

From a quantitative point of view, the amount of blockers or helpers might contribute in determining the difficulty in some way, as we have already observed. But there is an important aspect that cannot be overlooked in a game of this type, which affects not only helpers or blockers, but any other element of the board: the distribution of the elements. As a dummy example, consider Figs. 3.4 and 3.5, which could represent two different levels, both with the same number of blockers, but with clearly different level difficulty.

The most pure and complete way of including the distribution of elements to train any model would be to use directly the matrices that describe the position of each element on the board, but as we have said at the beginning, we cannot follow this approach.

We therefore make the following observation. Let us consider, for the moment, a level that only contains color bubbles. Remember that the way of destroying bubbles

Fig. 3.4 Level with 11 blockers (in black) which cannot be destroyed by hitting them. This level is impossible to complete (without using the orb)

Fig. 3.5 Level with 11 blockers (in black) which cannot be destroyed by hitting them. Since they are all on top of the board, they have no effect at all

on the screen is by shooting color bubbles to the ones that appear on the board and that match the color of the one we are shooting. When this happens, not only the bubble we hit is destroyed, but also any other bubble of the same color that is connected with it, either directly or via a path of bubbles of that color. In fact, individual, isolated bubbles cannot be destroyed with one single shot.

Therefore, we realize that, rather than the position of the color bubbles on the board, it is the distribution and size of *clusters of color bubbles* which seems to be important in the dynamics of the game. A few big clusters of color bubbles are easier to clear from the board, and thus require less shots, than lots of small clusters. In fact, if there was no gravity and we assumed that clusters of the same color are at least separated by two bubbles (to prevent that they become one single cluster when connected via a bubble of that color shot by the player), we would actually have that the number of required shots in order to complete the level would be bounded by below by the total number of color clusters.

Gravity makes it easier, but we can only take advantage of it if there are big holes. Thus, distribution of clusters is important, and also how they are connected. Similar things affect to clusters of other types, like blockers and helpers.

Regarding modifiers, since their effect is only appreciable when interacting with the bubbles they modify, we might be interested in knowing densities of modifiers on each cluster. For instance, a big cluster with a high density of fairies would make everything easier, because destroying that cluster would unleash an army of little gremlins that would destroy everything they find.

3.5 Improved Queue Sampling Using Bayesian Methods

The last part of the work is based on the following facts:

- The data is high dimensional and extremely complicated; models are unlikely to be able to capture all effects entirely (especially with small datasets).
- It is likely that the difficulty level will always have to be tested, and tweaked.
- Currently, tweaking is done by seeing how long it takes people to play the game, and adding or reducing the queued balls to achieve optimal difficulty.
- These methods will not require any understanding of the complex relationship that maps the feature space onto the difficulty.

Therefore, one aims to deduce the optimal ball queue for any given difficulty by performing more effective sampling methods. Bayesian Inference techniques specify how you should update your beliefs upon seeing data:

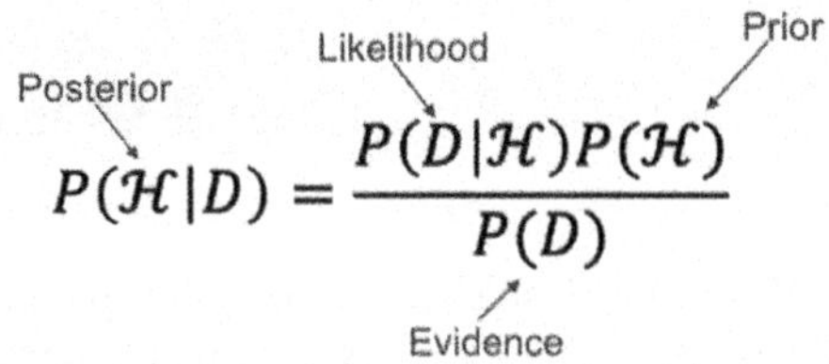

We shall use this to estimate the mean number of times it will take someone to complete a level given a number of queued balls. In particular, we assume that for a given number of queued balls, the number of attempts before completion of the level follows a geometric distribution,

$$Y_i \sim \text{Geom}(\theta). \tag{3.1}$$

We wish to find the θ associated with the number of balls of the queue, i, since $1/\theta$ represents the mean number of attempts before success, which is our quantity of interest.

Now, in order to perform updates on the prior we suggest a different experiment design than the one that is currently used by the company, so we shall proceed as follows. We let $\pi(\theta)$ denote our prior about the difficulty of the level. Then,

- if we have no real information the prior is assumed to be uniform,
- a more sophisticated prior should be considered in case of having some existing knowledge about the setup.

Now assume we give the level to someone who completes the game in time X_1, then we can update our prior as follows

$$p(\theta|X_1) = \frac{p(X_1|\theta)\pi(\theta)}{p(X_1)}. \tag{3.2}$$

Now assume we had n people playing the game, and so receive completion times $\{X_1, ..., X_n\}$, updating becomes as follows:

$$p(\theta|X_1, ..., X_n) = \frac{p(X_1, ..., X_n|\theta)\pi(\theta)}{p(X_1, ..., X_n)}. \tag{3.3}$$

Assuming independence between people makes this quantity simple to compute, and we now have a new estimate for the parameters of our distribution. Another important variable to quantify the level difficulty is the queue length. In particular, it is interesting to determine the queue length that takes (on average) the optimal number of attempts before completion. To update the scheme in order to select the queue length with the optimal completion time, one may proceed as follows:

1. Consider only queue lengths in a feasible region.
2. Set-up a discretised uniform prior for each different number of attempts (queue length).
3. Give the level with different numbers of queued balls to different people and perform bayesian updating. This can be iterated to a desired posterior probability.

We now know how to approximate the geometric distribution for each ball. This requires information on the time it takes for a player to pass to the next level for a given number of queued balls. We wish to devise the optimal number of balls with as little data as possible. The question is how can we use as small a number of samples as possible to converge to the correct solution?

Giving people levels with all possible queue lengths does not seem a good strategy. It is better to test more *queue* samples with queue lengths that we are more confident fit our hypothesis (so we can either accept or reject). The idea is then to use our current posterior estimate to speed up this process by optimising the queue lengths that we give to users.

A Particular Example Assuming Player Learning

To illustrate the ideas above we now consider the following particular example:

- A new level has been designed and one wants to determine its level of difficulty.
- It has been played by designers who found: (1) The level was easy to complete with 12 queued shots. (2) It was very difficult (but possible) to complete with 5 queued shots.
- 6 attempts have been deemed to be the most appropriate number of levels for user retention.
- **The Goal** is to find (as fast as possible) the number of queued balls that gives the closest to an average of 6 level attempts.

Thus far, we have assumed that the probability of passing a level is independent of the attempt number. However, in reality, players are likely to learn how to pass the level each time they play, and thus the probability of passing a level p_i on the n_{th} attempt would be a monotonic increasing function of n. In Fig. 3.6-left we show an example of how p_i could depend on n.

With the assumption that $p_i = p_i(n)$, X_i would no longer follow a geometric distribution. Using the function $p_i(n)$ from Fig. 3.6-left, in Fig. 3.6-right we show how the probability of passing a level *for the first time* would vary as the number of attempts increases in this example. We observe that assuming that players learn, it is most likely that players would pass for the first time on their second attempt, not their first as it is implicitly implied when assumed a fixed probability p_i independent of n.

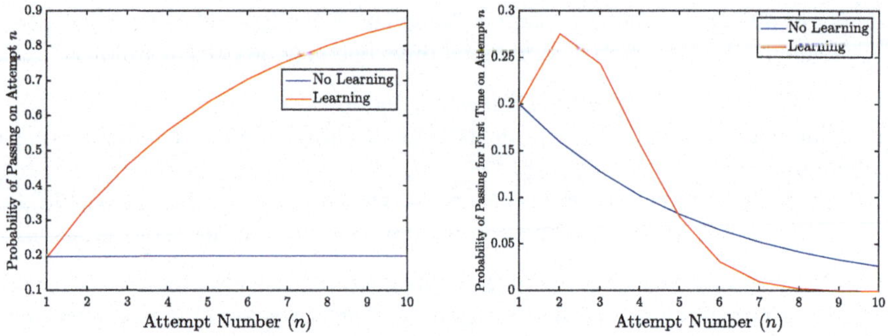

Fig. 3.6 Left: Probability of passing a level as the attempt number increases. Right: Probability of passing a level for the first time as the attempt number increases

3.6 Discussion and Causal Relationships

The ultimate objective is to create a function $Y = f(X)$ predicting the level of difficulty by means of a set of features describing the design of the level. Since it is different predicting than explaining, we must be aware of the final use of this model in the way we create it.

The main assumption of a predictive model is that data is independent and identically distributed, which implicitly implies that we are assuming that the past and the future behave in the same way. However, when we wonder how the level of difficulty will change when we modify some element of the game design, we are potentially asking about a behaviour that does not reflect past dynamics. These types of questions are studied by causal inference techniques [7].

In the process of creating a function $Y = f(X)$ we may need to summarise or to transform the set of features X, due to the high-dimensionality of this problem. The set of features is strongly correlated, which means that omitting a relevant variable may not be a problem for predicting purposes, since statistically the information of the missing variable can be recovered by the rest.

As an example, imagine that one wants to describe the level of difficulty of a particular level in terms of the *queue* length (number of available moves). It may happen that historically, due to how designers created the levels, all levels with high difficulty have a small queue length. Moreover, in order to compensate such difficulty, the designer at the same time increased the number of helpers. In this way, the variables *queue* and *helpers* are highly and negatively correlated.

There is not a large difference in terms of accuracy when training a machine learning model using only the *queue* variable or both variables at the same time. Since they are highly correlated, if you know the queue length, you may also know with high precision the helpers number, and thus the latter brings little extra information to the model.

How a model trained only using the *queue* variable will predict a new design with a low queue length but a low number of helpers? This is a new situation very

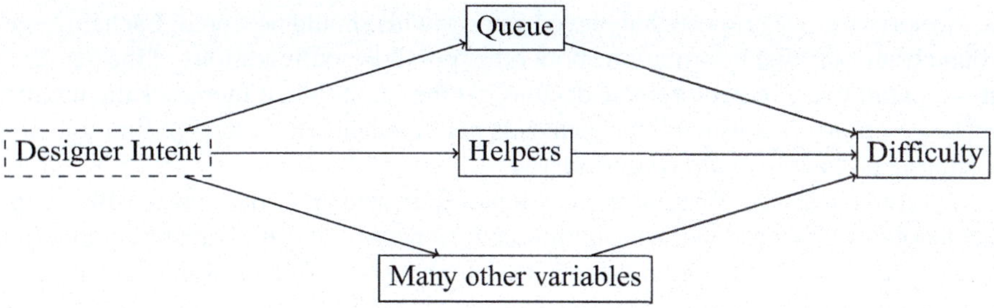

Fig. 3.7 Design dynamic

different from the past and since we have not included the *helpers* variable in the model, our model will fail to infer the level difficulty.

It seems natural that if one misses a particular variable, it will not be possible to make good predictions about it in the future. However it is more problematic than that. The estimated effect of the remaining variable, *queue*, may neither be estimated correctly. Imagine again the situation where *queue* and *helpers* are highly correlated, but in this case the real source of difficulty is the number of helpers, while *queue* has little or no effect at all. The queue length itself is a very good predictor of the difficulty, so the model will put a lot of weight into it, the main reason being the lack of the actual relevant variable, the *helpers* variable. But we know that, in fact, *queue* should have no relevance in the model.

Methodologically speaking, there will be two models which differ in just one feature. Both with a very similar level of accuracy, but making totally different predictions. This implies that the usual metric to evaluate models, the cross-validated accuracy or any of its variants based on cross validation, may not be reliable for this particular problem.

The example above is very simplistic. In reality, designers will make a large number of decisions. Those are taken based on their intent. In this case we say that the *designers' intent* is a confounding variable. Such variables are not observable nor can they be easily described. Fortunately the intent is ultimately reflected by the game design, so including the whole set of variables that completely defines the game is enough to avoid any of the problems expressed above. This can be deduced from Pearl's backdoor criterion. Figure 3.7 represents this design dynamic.

One could be tempted to choose a subset of variables suspected of being the most relevant ones. But this comes at a risk, as explained above, that cannot be measured in terms of cross validation predictive accuracy. In fact, each designer may have their own preference for some geometric patterns that may potentially have an impact on the game's difficulty. Thus determining which are the relevant features can be a hard problem.

In order to be sure no relevant information is missing, one can go for the most simple answer: include them all. But this leads to a statistical problem, namely, the high dimensionality, since we have many more variables than observations.

To solve this we propose that played game rounds should be stored. Each different round brings its own information about other possible configurations of the game, so it is a natural way to explore new designs. At the same time it increases the number of observations. If storing all the rounds is not practical due to storage limitations, a sampling procedure could be considered.

Avoiding any type of variable selection or dimensionality reduction, with a large database of played rounds, we can train a deep neural network capable of handling large volumes of data. The obtained model could be safely used to make reliable predictions for new designs.

References

1. Individual differences in the proneness to have flow experiences are linked to dopamine d2-receptor availability in the dorsal striatum. NeuroImage 67:1–6 (2013)
2. M. Csikszentmihalyi, *Flow: The Psychology of Optimal Experience* 01 (1990)
3. A. Dietrich, Neurocognitive mechanisms underlying the experience of flow. Consciousness Cognit. **13**(4), 746–761 (2004)
4. R. Hunicke, M. Leblanc, R. Zubek, Mda: A formal approach to game design and game research, in *Proceedings of the Challenges in Games AI Workshop, Nineteenth National Conference of Artificial Intelligence* (Press, 2004), pp. 1–5
5. C. Kilgore, Gamasutra (2020). https://www.gamedeveloper.com/design/understanding-challenge
6. C.N. Morris, Natural exponential families with quadratic variance functions: statistical theory. Ann. Stat. **11**(2), 515–529, 06 (1983)
7. J. Pearl, *Causality: Models, Reasoning and Inference*, 2nd edn. (Cambridge University Press, USA, 2009)
8. Statista. Statista (2020). www.statista.com/statistics/237749/value-of-the-global-entertainment-and-media-market/

Chapter 4
Mathematical Modelling of Fibre Coating

Maria Aguareles, Francesc Font, Tim Myers, and Jordi Ripoll

Statement of the Problem

The research department at Frenos Sauleda S.A. is trying to find an alternative formulation for one of his produced materials. There are two components in its formula, toluene and sepiolite, which are yet forbidden in some countries and could be forbidden shortly also in Spain.

Toluene is a volatile organic compound used commonly as a solvent. It may have local as well as systemic harmful effects. It may cause irritation of the eyes, respiratory tract, and skin. Other symptoms include headache, dizziness, fatigue, muscular weakness, collapse, and coma. Sepiolite is an abundant fibrous clay mineral with modest to good adsorption properties and extensive industrial process applications. The applications are based on following three types of properties: adsorptive, rheological, and catalytic properties. The typical industrial applications are decoloring agents, absorbent granules, cat litter, carriers for insecticides and herbicides, dispersants, rubber and plastic industries, asbestos substitution and cosmetics. Sepiolite is produced commercially almost exclusively in Spain. It is currently believed that fibrous minerals are potentially active to induce mesothelioma. Their carcinogenic effects are thought to be related to fiber size and shape, rather than chemical composition. They appear to behave in biological systems in much the same way as the asbestos minerals.

The product to develop is formed by thread and paste. Toluene and sepiolite are the components of the paste to be replaced. The thread-paste ratio must be maintained. The union of the thread and the paste is done through an extrusion process. The new components and proportions of the formula must produce a paste that behaves

M. Aguareles · J. Ripoll
Department d'Informatica, Matematica Aplicada i Estadistica, Universitat de Girona, Girona, Spain

F. Font (✉) · T. Myers
Centre de Recerca Matemàtica, Campus de Bellaterra Edifici C, 08193 Bellaterra, Barcelona, Spain
e-mail: francesc.font@upc.edu

M. Aguareles et al. (eds.), *Applications of Industrial Mathematics*,
RSME Springer Series 8, https://doi.org/10.1007/978-3-031-32130-6_4

similarly to the current one in terms of wet and flowing, to keep the extrusion machine working properly. Toluene reduces the viscosity of the paste, it keeps the material flowing, we cannot. We need to replace it with a liquid resin of similar viscosity. We know the viscosity of the raw materials, but we don't have a viscometer to check the viscosity of the mix.

The control variables in the process are the following:

- Spindle speed and thread entry speed (they depend on the machine voltage selected).
- head temperature (melting temperature).
- Product weight per meter.
- Formulation.

We must get a compromise between all these variables. Low speed reduces production, high temperature affects machine performance and lifespan. We need to keep also the performance properties of the final product. The control properties of the product are coefficient of friction, wear and density.

We look for directions on how to set up and organize our essays in the most optimal way.

4.1 Introduction

It is anticipated that in the near future toxic components, such as toluene and sepiolite, will be prohibited in the manufacturing process used by Frenos Sauleda S.A to fabricate clutch components. Toluene is already listed as a restricted substance by the European Chemicals Agency [1] and sepiolite, although not restricted by ECA yet, can be potentially carcinogenic depending on the characteristic size of its fibres [2]. The study group was asked to investigate the process and come up with recommendations to ensure the quality of the final product after the removal of these components.

A key stage in the process is the coating of wires via extrusion. Three wires enter the extruder, these are then entangled and coated to form a single component. Upon leaving the extruder the composite wire should have a specified coating thickness and/or weight. An illustration of a typical wire coating via extrusion process is presented in Fig. 4.1. During the meeting, work focussed on this wire coating process.

The coating material starts as a wet powder which is fed into a hopper and subsequently heated so that certain components melt, resulting in a thick paste similar to plasticine. The paste is composed of a number of components, including plastic. This together with the sepiolite suggests that the paste will have significant non-Newtonian behaviour. A Newtonian fluid exhibits a constant viscosity when subjected to a shear stress, non-Newtonian fluids can exhibit many unusual behaviours such as shear thinning or thickening, elasticity or a yield stress (below this they behave as a solid, above as a liquid).

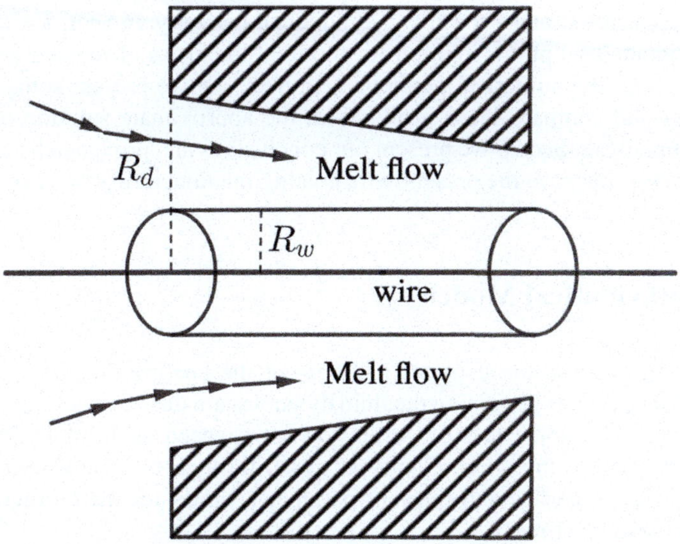

Fig. 4.1 Schematic of wire coating process

The role of the toluene, a volatile solvent, is to reduce the viscosity during the coating process but then rapidly evaporate over a very short distance (of the order of centimetres) after leaving the extruder. Sepiolite is a fibrous clay mineral with good adsorbent properties. The fibrous structure can add strength to a solid material, in a liquid it can result in thixotropic (time-dependent shear thinning) behaviour.

The process of coating a solid surface with a fluid is relevant in many industrial applications. The process typically involves the withdrawal of a solid from a liquid bath, and is sometimes referred to as *the drag-out problem*. The key aspect in the drag-out problem is to predict the final height of the coating film and find out the most relevant physical parameters involved in the process. There exists a number of interesting reviews on the specific topic of coating [3, 4] and, in more general terms, on thin liquid films [5, 6]. Nowadays, there are whole scientific journals dedicated to coating research [7, 8]. Probably the first theoretical studies describing the coating process were carried out in the 1940s by Derjaguin [9] and Landau and Levich [10], in an attempt to mathematically model the deposition of a photosensitive emulsion on a solid surface for cinefilm manufacture [10]. An extension of the theory was introduced by White and Tallmadge [11] which allowed the prediction of the evolution of the film thickness for a wider variety of coating fluids. Later, the drag out problem was analysed in more detail and solutions to the governing equations were obtained by means of matched asymptotic expansions [12–14].

In this report, in order to predict the final thickness of the coating used in Frenos Sauleda wires we will take an approach similar to that in Landau and Levich [10]. The outline of the report is as follows. In Sect. 4.2 we describe the main assumptions we make to describe the coating process and formulate the mathematical model. In Sect.

4.3, we discuss the strategy to solve the model numerically. In Sect. 4.4, we provide an approximate asymptotic solution for the final thickness of the coating film and compare it with the numerical prediction. In Sect. 4.5 we present some numerical simulations and compare these results with the approximate solution obtained in Sect. 4.4. Finally, in Sect. 5 we present our conclusions and provide a list of possible future routes to improve the present wire coating manufacturing process.

4.2 Mathematical Model

When the wire leaves the extruder it drags-out the coating fluid as illustrated in Fig. 4.2. The equations governing the fluid dynamics are the Navier-Stokes equations which are to be coupled to surface tension experienced at the fluid free surface ($y = h(x, t)$) and to the wire velocity at $y = 0$ through the boundary conditions. Some material properties and relevant parameter values for the current industrial process are listed in Table 4.1.

Solving the Navier-Stokes equations in their most general form is very challenging. In order to reduce the complexity of the problem we make the following assumptions:

(a) We assume that the radial curvature plays a negligible role in determining the film shape. Consequently the film shape varies only with x and t (it is demonstrated in Appendix B that this leads to errors of the order 1%). This allows us to define a vertical co-ordinate y measured from the fibre surface.
(b) The coating behaves as a Newtonian incompressible fluid.
(c) Gravity effects are negligible.
(d) Since the ratio of the initial film thickness to the fibre radius is small (approximately 0.07), the coating layer can be considered a thin film.

After these assumptions are made the Navier-Stokes equations are substantially reduced (see Eqs. (4.18)–(4.19) in Appendix A) and an expression for the fluid velocity can be obtained (see Eq. (4.24) in Appendix A).

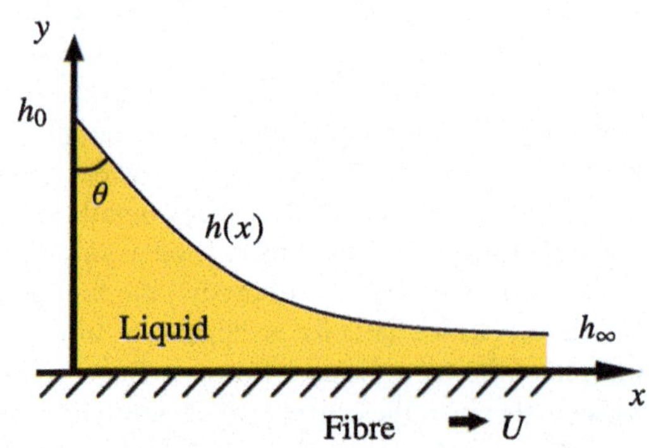

Fig. 4.2 Sketch of the fluid exiting the extruder

Table 4.1 Typical parameter values required for the model

Quantity	Symbol	Value	Units
Liquid density	ρ	2000	kg.m^{-3}
Liquid viscosity	μ	10	Pa . s
Surface tension	σ	$30{\cdot}10^{-3}$	N.m^{-1}
Gravity	g	9.81	m.s^{-2}
Fibre velocity	U	2	m.s^{-1}
Fibre radius	R	0.003	m
Initial film thickness	h_0	0.0002	m

On a flat surface the drag-out problem is governed by a standard mass balance:

$$\frac{\partial h}{\partial t} + \frac{\partial Q}{\partial x} = 0\,, \tag{4.1}$$

where, as depicted in Fig. 4.2, $h(x,t)$ is the height of the liquid layer and $Q(x,t)$ is the fluid flux per unit length. A derivation of equation (4.1) can be found in Appendix A. Following the derivation and the notation used in the Appendix we find the flux is given by

$$Q = \int_0^h u\, dy = \frac{\sigma}{3\mu} h^3 h_{xxx} + Uh\,, \tag{4.2}$$

where σ is the air-liquid surface tension, μ the dynamic viscosity, U the speed of the fibre, u the fluid velocity, and the subscript x indicates the derivative of $h(x,t)$ with respect to x. Hence

$$\frac{\partial h}{\partial t} + \frac{\partial}{\partial x}\left(\frac{\sigma}{3\mu} h_{xxx} h^3 + Uh\right) = 0\,. \tag{4.3}$$

This equation shows that the height of the fluid layer varies due to the effect of surface tension (more precisely the ratio of surface tension to viscosity) and the pulling speed.

Most drag-out processes are continuous, running at a constant speed. Mathematically this means we may treat the process as steady-state, hence we neglect the time derivative in both (4.1) and (4.3) to find an ordinary differential equation for the flux. If we note that far from the outlet the height is constant and then denote it by h_∞ then the far-field flux is $Q = Uh_\infty$ and integrating the steady form (i.e. $\partial h/\partial t = 0$) of (4.3) leads to

$$h_{xxx} = \frac{3\mu U}{\sigma} \frac{(h_\infty - h)}{h^3}\,. \tag{4.4}$$

This is a standard form of equation, applicable to a multitude of physical processes. Various forms, both steady and unsteady, are discussed in [6]. The (non-dimensional) ratio $\mu U/\sigma$ is known as the capillary number and represents the relative importance of viscous drag to surface tension forces. Since Eq. (4.4) is third-order we expect three boundary conditions, however it is autonomous (invariant under shifts in x) which reduces the required conditions to two. Below we write down three but the h_∞ condition has already been imposed to obtain (4.4)

$$\lim_{x\to\infty} h = h_\infty, \qquad h(0) = h_0, \qquad \frac{d}{dx}h(0) = -\cot\theta\,, \tag{4.5}$$

where h_0 and θ are the thickness and exit angle of the fluid as it leaves the extruder as shown in Fig. 4.2.

Dimensionless Model

Mathematically it is simplest to work in terms of a dimensionless model. We denote $\hat{x} = x/L$ and $\hat{h} = h/h_\infty$. The height-scale h_∞ is chosen as the final thickness, the length-scale is as yet unknown. The governing equation (4.4) becomes

$$\frac{h_\infty}{L^3}\hat{h}_{\hat{x}\hat{x}\hat{x}} = \frac{3\mu U}{\sigma h_\infty^2}\frac{(1-\hat{h})}{\hat{h}^3}\,,$$

which indicates the length-scale $L = h_\infty\sqrt[3]{\sigma/(3\mu U)}$. The governing equation and boundary conditions may now be written

$$\hat{h}_{\hat{x}\hat{x}\hat{x}} = \frac{1-\hat{h}}{\hat{h}^3}\,, \tag{4.6}$$

which we solve subject to

$$\lim_{\hat{x}\to\infty}\hat{h}(\hat{x}) = 1 \qquad \hat{h}(0) = \hat{h}_0 = \frac{h_0}{h_\infty} \qquad \hat{h}_{\hat{x}}(0) = -\frac{L}{h_\infty}\cot\theta = -\cot\theta\sqrt[3]{\frac{\sigma}{3\mu U}}\,. \tag{4.7}$$

An important feature of the non-dimensional formulation is that it clearly shows the dependence of the problem on the physical conditions. Now we can see that there are only two important parameter values $\hat{h}_0 = h_0/h_\infty$ and $\sqrt[3]{\sigma/(3\mu U)}\cot\theta$. The physical meaning of this is that we may operate at different conditions but achieve the same final film height if we keep these parameters constant. This is key to the present study and indicates how to compensate for process changes. For example, reducing toluene leads to an increase in viscosity, the same final thickness may then be achieved by a corresponding reduction in fibre speed U or an increase in surface tension. The contact angle θ clearly plays an important role in the coating thickness, this may be varied by changes in outlet shape, fibre speed or through the addition of surfactants.

4.3 Numerical Solution

To obtain a more complete picture of the process we now examine the numerical solution of the system.

The numerical problem is easily obtained using a marching scheme and shooting (see [15]). The starting point is based on the fact that sufficiently far from the outlet the height is approximately h_∞, that is

$$\lim_{\hat{x}\to\infty} \hat{h}(\hat{x}) = 1 \tag{4.8}$$

so we may write $\hat{h}(\hat{x}) = 1 + \epsilon f(\hat{x})$ where $\epsilon \ll 1$ and $f(\hat{x})$ is an unknown function. Substituting into the governing equation (4.6)

$$\epsilon f_{\hat{x}\hat{x}\hat{x}} = \frac{1-(1+\epsilon f)}{(1+\epsilon f)^3} = -\epsilon f(1 - 3\epsilon f + \cdots)\,. \tag{4.9}$$

We can choose ϵ arbitrarily small, provided we move sufficiently far from the outlet, which also corresponds to linearising equation (4.6) about the constant solution $\hat{h} \equiv 1$. Hence we may neglect terms of order ϵ and find

$$f_{\hat{x}\hat{x}\hat{x}} = -f\,. \tag{4.10}$$

This equation has three solutions of the form $f = e^{m\hat{x}}$, where $m^3 = -1$. There is one real solution $m = -1$ and two complex ones. The complex solutions lead to an oscillating film. Physically this occurs when the coated wire enters into the outlet rather than leaving it (i.e. the velocity U is negative). In the present case we deal with the drag out problem, so the film decreases smoothly in height, hence only the solution with $m = -1$ is physical. Consequently, far from the outlet the height varies according to

$$\hat{h}(\hat{x};\epsilon) = 1 + \epsilon e^{-\hat{x}}\,. \tag{4.11}$$

This is the form we use to start the marching scheme. We therefore solve equation (4.6) in a bounded domain $\hat{x} \in [0, \ell]$, with initial conditions

$$\hat{h}(\ell;\epsilon) = 1 + \epsilon e^{-\ell}, \quad \hat{h}_x(\ell;\epsilon) = -\epsilon e^{-\ell}, \quad \tilde{h}_{xx}(\ell;\epsilon) = \epsilon e^{-\ell}\,. \tag{4.12}$$

In practice we fix $\ell \gg 1$ and choose ϵ so that $\epsilon e^{-\ell}$ is small. We then use $x = \ell$ as the initial point from which we march backwards until $x = 0$. For a fixed ℓ, the idea is therefore to find a specific $\epsilon = \epsilon^*$ such that

$$\hat{h}_{\hat{x}}(0; \epsilon^*) = -\cot\theta\sqrt{\frac{\sigma}{3\mu U}} . \tag{4.13}$$

We note that the gradient boundary condition depends only on physical parameters of the problem so (4.13) provides an equation for ϵ which we solve using the Newton-Raphson method. Finally, to obtain h_∞ we note that h_0 is the initial film thickness, which is a known physical quantity, and therefore

$$h_\infty = h_0/\hat{h}_0 , \tag{4.14}$$

where $\hat{h}_0 = \hat{h}(0; \epsilon^*)$. In Sect. 4.5 we present some results obtained from the numerical simulations.

4.4 Approximate Solution

Returning to the approximate solution employed to start the marching scheme we assume that there is only a small change in height from the outlet to the far-field. In which case the approximate solution may be applied throughout the domain. Equating the approximate solution, Eq. (4.11), with the angle at $\hat{x} = 0$ then determines $\epsilon = L\cot\theta/h_\infty$ and so

$$\hat{h}(\hat{x}) = 1 + \frac{L}{h_\infty}\cot\theta\, e^{-\hat{x}} . \tag{4.15}$$

At $\hat{x} = 0$ we find

$$\hat{h}_0 = 1 + \sqrt[3]{\frac{\sigma}{3\mu U}}\cot\theta . \tag{4.16}$$

Returning to dimensional form and rearranging gives an expression for the final film height

$$h_\infty = \frac{h_0}{1 - \sqrt[3]{\sigma/(3\mu U)}\cot\theta} . \tag{4.17}$$

In the following section we will show that this approximate solution provides good agreement with the values obtained through direct numerical simulations.

4.5 Results

As explained in Sect. 4.3, we start our shooting scheme by fixing ℓ and an initial value for the shooting parameter, $\epsilon = \epsilon_0$, which, by marching backwards from $x = \ell$ with (4.12) as initial conditions, provides an initial profile for $\hat{h}_0(\hat{x}; \epsilon_0)$. Newton-Raphson's method then provides the value of ϵ^* such that $\hat{h}_{\hat{x}}(0; \epsilon^*)$ achieves the prescribed value given in (4.13). As a way of illustration, in Fig. 4.3 we fix $\ell = 10$ and show in the same graph the initial profile, $\hat{h}_0(\hat{x}; \epsilon_0)$ and the final one $\hat{h}_0(\hat{x}; \epsilon^*)$, where the gradient boundary conditions at $\hat{x} = 0$ is satisfied.

For each choice of $\hat{h}_{\hat{x}}(0)$ a different initial height $\hat{h}_0$ is obtained. To ensure realistic values for $\hat{h}_{\hat{x}}(0)$ we consider the parameter values shown in Table 4.1 and substitute in Eq. (4.13) considering angles θ ranging from $\pi/8$ to $3\pi/8$. The results are shown in Fig. 4.4.

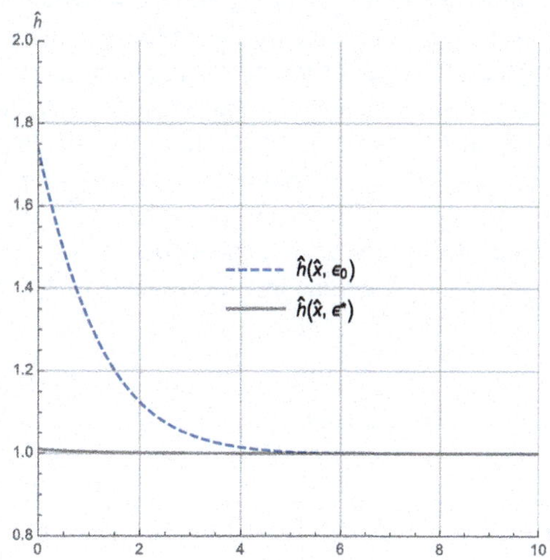

Fig. 4.3 $\hat{h}_0(\hat{x}; \epsilon_0)$ and $\hat{h}_0(\hat{x}; \epsilon^*)$ for $\ell = 10$, $\epsilon_0 = 1$, $\epsilon^* = 0.009774$ for $\hat{h}_0'(0; \epsilon^*) = -0.01$

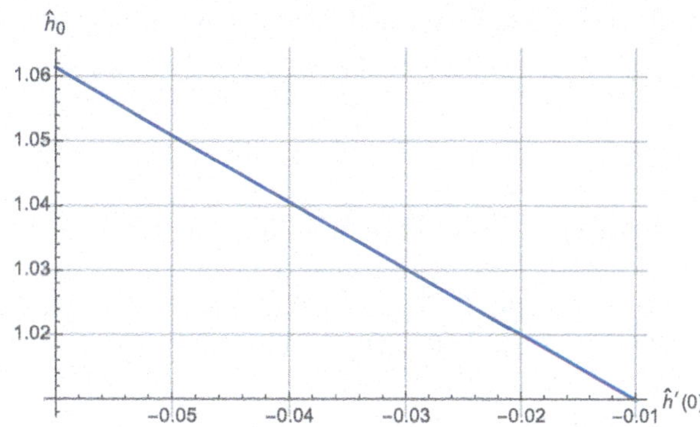

Fig. 4.4 $\hat{h}_0'(\hat{x}; \epsilon_0)$ versus $\hat{h}_0(\hat{x})$

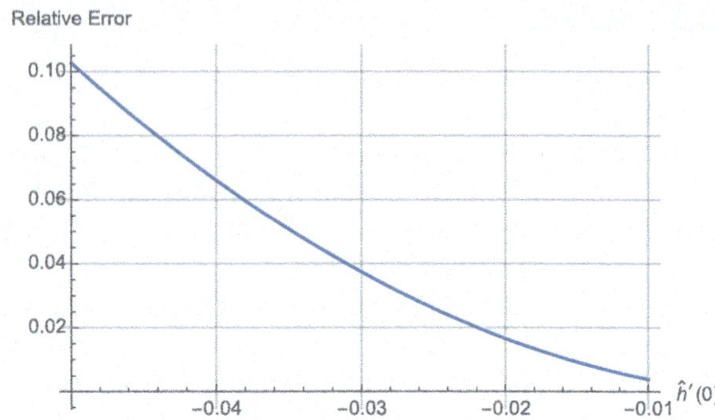

Fig. 4.5 Relative error (in percentage form) for the values of h_∞ obtained with the approximate expression (4.17) with respect to the numerically obtained one for different initial gradient values $\hat{h}_{0\hat{x}}(0)$

In terms of applicability to an industrial setting, it is obvious that the approximate expression (4.17) would be more useful than having to go through numerical simulations to compute h_∞. In Fig. 4.5 we show the relative error between the approximate formula and the numerically obtained values. The error decreases with the initial angle: for values of the initial gradient $\hat{h}_{0\hat{x}}(0)$ between -0.05 and -0.01 (which correspond to exit angles from $3\pi/8$ to $\pi/8$) the maximum relative error is around 10% and it takes place for the angle of $3\pi/8$ which, according to the sketch in Fig. 4.2, would correspond to a fluid exiting the extruder almost horizontally. With $\hat{h}_{0\hat{x}}(0) \approx -0.01$ the error is less than 1%. This results suggests that for realistic exit angles the approximate expression for the final film height, Eq. (4.17), may be used to accurately estimate the dependence of h_∞ on the operating conditions.

4.6 Conclusion

In this report we have developed a basic model to determine the variation of the thickness of the paste as it emerges from the extruder and is carried along with the wire. It is a simple matter to solve the equation numerically, however the approximate analytical solution shows that the final paste height is well approximated by

$$h_\infty \approx \frac{h_0}{1 + \sqrt[3]{\frac{\sigma}{3\mu U}} \cot\theta} .$$

From this we can see the controlling parameters for the process: the initial height, surface tension, viscosity, velocity and angle of exit at the extruder. The final component, $\cot\theta$, depends on the internal dynamics of the extruder and so are beyond this study. The other parameters are simpler to control. The final product weight may be estimated by assuming the three wires used in the process may be represented by an equivalent wire of radius of R, then

$$W \approx \rho_p g \pi \left[(R + h_\infty)^2 - 3R_w^2\right] + 3\rho_w g \pi R_w^2 \,, \qquad R \approx 2R_w \,,$$

where R_w is the actual wire radius.

The analytical solutions permit various observations, specifically we can see how to retain a layer of the same height/weight by adjusting different parameters. For example, if we replace toluene resulting in a decrease in μ but keep σ fixed we can still achieve the same final thickness by increasing U accordingly. If the toluene is all evaporated its replacement by another volatile solvent will not affect the final weight. On the other hand sepiolite remains in the final product so a replacement should be of a similar density.

Future work:

For the academics

- Extend the model to include variable fluid properties (for example the paste is non-Newtonian);
- Include evaporation (preliminary calculations suggest a 13% decrease in height);
- Investigate the existing wide literature on wire extrusion, including non-Newtonian effects.

For Frenos Sauleda

- Carry out detailed experiments to determine the properties of the melted paste and also to measure the angle θ, which is crucial to the results and depends on paste and flow conditions.

Appendix A: General Equation for the Film Height

For the steady flow of a thin layer of fluid being dragged along on a fibre the dominant terms in the Navier-Stokes equations are

$$0 = -\frac{\partial p}{\partial x} + \mu \frac{\partial^2 u}{\partial y^2} \tag{4.18}$$

$$0 = -\frac{\partial p}{\partial y} \,, \tag{4.19}$$

where $p(x, y)$ is the pressure and $u(x, y)$ is the horizontal component of the velocity at each point (x, y) of the fluid. These require solving subject to the following boundary conditions

$$u = U \,, \quad v = 0 \quad \text{at } y = 0 \tag{4.20}$$

$$\frac{\partial u}{\partial y} = 0 \,, \quad p = -\sigma\kappa \quad \text{at } y = h(x, t) \,. \tag{4.21}$$

On the surface $y = 0$ the fluid in contact with the fibre matches the fibre velocity U, with no vertical velocity. At the free surface $y = h(x, t)$ there is no sheer stress and the vertical stress is due to surface tension, where σ is the surface tension and the curvature

$$\kappa = \frac{h_{xx}}{\left(1 + h_x^2\right)^{3/2}} \tag{4.22}$$

where the subscript in h_x indicates derivative with respect to x. Equations (4.18, 4.19) are consistent with lubrication theory, which is based on the the assumption the height is significantly less than the length, in line with this we find $\kappa \approx h_{xx}$.

Equation (4.18) indicates $p = p(x)$ is a function of x alone. The boundary condition then indicates $p = \sigma\kappa$ everywhere. Equation (4.19) then becomes

$$\sigma h_{xxx} + \mu \frac{\partial^2 u}{\partial y^2} = 0 . \tag{4.23}$$

Integrating (4.23) twice and applying the boundary conditions (4.20)–(4.21) we obtain

$$u = \frac{\sigma}{2\mu} h_{xxx}\, y(2h - y) + U . \tag{4.24}$$

The kinematic condition

$$\frac{d}{dt}(y(t) - h(x(t), t)) = 0 \implies v(h) = u(h)\frac{\partial h}{\partial t} + h_x , \tag{4.25}$$

is the mathematical statement that fluid particles on the surface stay there. For an incompressible fluid

$$\frac{\partial u}{\partial x} + \frac{\partial v}{\partial y} = 0 . \tag{4.26}$$

We may integrate this between $y = [0, h]$

$$v(h) = -\int_0^h \frac{\partial u}{\partial x}\, dy \tag{4.27}$$

after noting $v(0) = 0$. Combining this with the kinematic condition

$$u(h)h_x + \frac{\partial h}{\partial t} = -\int_0^{h(x)} \frac{\partial u}{\partial x}\, dy . \tag{4.28}$$

Leibniz theorem shows that

$$\frac{\partial}{\partial x}\int_0^h u\, dy = \int_0^h \frac{\partial u}{\partial x}\, dy + u(h)h_x x . \tag{4.29}$$

Hence

$$u(h)\frac{\partial h}{\partial x}+\frac{\partial h}{\partial t}=-\left[\frac{\partial}{\partial x}\int_0^h u\,dy-u(h)h_x\right]. \tag{4.30}$$

Defining the fluid flux

$$Q=\int_0^{h(x)} u\,dy=\frac{\sigma}{3\mu}h_{xxx}h^3+UH \tag{4.31}$$

we obtain the well-known mass balance

$$\frac{\partial h}{\partial t}+\frac{\partial Q}{\partial x}=0. \tag{4.32}$$

Then, taking the derivative with respect to x of (4.31) gives

$$\frac{\partial Q}{\partial x}=\int_0^{h(x)}\frac{\partial u}{\partial x}dy+h_x u \tag{4.33}$$

Finally, combining (4.28) and (4.33) we obtain the general equation for the film hight

$$\frac{\partial h}{\partial t}+\frac{\partial Q}{\partial x}=0. \tag{4.34}$$

Appendix B: Flat or Curved Surface?

Obviously the geometry of the problem involves radial symmetry, indicating we should analyse the governing equations in a radial system. However when the liquid layer is thin it is often possible to work in a Cartesian system. To see whether this is justified in the present situation we consider the stress on the paste surface. The stress is proportional to the curvature and in polar co-ordinates is given by

$$\kappa=\frac{(R+h)_{xx}}{(1+(R+h)_x^2)^{3/2}}+\frac{1}{R(1+(R+h)_x^2)^{1/2}}.$$

The first term on the right corresponds to the height variation along the axis, the second term is the circular stress. This is the stress that will pinch off a jet of water.The radius R is the constant radius of a wire, while h is the variable paste thickness, noting that R is constant we may write

$$\kappa=\frac{h_{xx}}{(1+h_x^2)^{3/2}}+\frac{1}{R(1+h_x^2)^{1/2}}\sim\frac{H}{L^2}+\frac{1}{R}\sim 10^5+10^3$$

The first term represents the stress that would occur on a film of height h on a flat surface, the second the circular stress. Given that the circular stress is 100 times smaller than the flat surface stress we may then ignore radial curvature (leading to errors of the order 1%). Consequently in this report we will deal only with the flat substrate equations.

Acknowledgements M.A., T.M., F.F. acknowledge the support of grant PID2020-115023RB-I00 financed by MCIN/AEI/ 10.13039/501100011033/ and by "ERDF A way of making Europe". T.M. acknowledges the CERCA Programme of the Generalitat de Catalunya, his work was also supported by the Spanish State Research Agency, through the Severo Ochoa and Maria de Maeztu Program for Centres and Units of Excellence in R&D (CEX2020-001084-M).

References

1. Substances restricted under REACH by European Chemicals Agency. https://echa.europa.eu/substances-restricted-under-reach, accessed: 5-09-2022
2. B. Bellmann, H. Muhle, H. Ernst, Investigations on health-related properties of two sepiolite samples. Environ. Health Perspect. **105**, 1049–1052 (1997). http://www.jstor.org/stable/3433508
3. D. Quéré, Fluid coating on a fiber. Annu. Rev. Fluid Mech. **31**(1), 347–384 (1999)
4. S.J. Weinstein, K.J. Ruschak, Coating flows. Annu. Rev. Fluid Mech. **36**(1), 29–53 (2004)
5. R.V. Craster, O.K. Matar, Dynamics and stability of thin liquid films. Rev. Mod. Phys. **81**, 1131–1198 (2009)
6. T. Myers, Thin films with high surface tension. SIAM Rev. **40**(3), 441–462 (1998)
7. Coatings, https://www.mdpi.com/journal/coatings, accessed: 14-05-2020
8. Surface and coating technology, https://www.journals.elsevier.com/surface-and-coatings-technology, accessed: 5-09-2022
9. B. Derjaguin, On the thickness of the liquid film adhering to the walls of a vessel after emptying. Progress Surface Sci. **43**(1), 134–137 (1993), http://www.sciencedirect.com/science/article/pii/007968169390022N
10. L. Landau, B. Levich, Dragging of a liquid by a moving plate, in *Dynamics of Curved Fronts*, ed. by P. Pelcé (Academic Press, San Diego, 1988), pp. 141–153. http://www.sciencedirect.com/science/article/pii/B9780080925233500162
11. D. White, J. Tallmadge, Theory of drag out of liquids on flat plates. Chem. Eng. Sci. **20**(1), 33–37 (1965), http://www.sciencedirect.com/science/article/pii/0009250965800410
12. D.F. James, The meniscus on the outside of a small circular cylinder. J. Fluid Mech. **63**(4), 657–664 (1974)
13. S.D.R. Wilson, The drag-out problem in film coating theory. J. Eng. Math. **16**, 209–221 (1982)
14. E.O. Tuck, L.W. Schwartz, A numerical and asymptotic study of some third-order ordinary differential equations relevant to draining and coating flows. SIAM Rev. **32**(3), 453–469 (1990)
15. Shooting method, Wikipedia, https://en.wikipedia.org/wiki/Shooting_method, accessed: 5-09-2022

List of participants

First name	Surname	Institute
Lluis	Abio	Frenos Sauleda, S.A.
Maria	Aguareles	Universitat de Girona
Jose	Antunes	Instituto Superior Técnico Lisboa
Thomas	Babb	University of Oxford
Llorenç	Badiella	Universitat Autònoma de Barcelona
Hakim	Bagua	University of Djelfa
Anas	Belfadil	Universitat Autònoma de Barcelona
Ernest	Benedito	Universitat Politècnica de Catalunya
Oliver	Bond	University of Oxford
Ferran	Brosa	University of Warwick
Joaquim	Bruna	Universitat Autònoma de Barcelona
Daniel	Castillo	Colles Castelleres de Catalunya
Mesut	Cevik	Kadir Has University
Arnau	Dòria	Universitat Politècnica de Catalunya
Ümit	Ertugrul	Karadeniz Technical University
Francesc	Font	Centre de Recerca Matemàtica
Xavier	Guardiola	King Shared Services
Javier J.	Gutiérrez	Universitat de Barcelona
Alex	Haro	Universitat de Barcelona
Adrián	Hinojosa	Universitat de Barcelona
Kieran	Jervis	University of Leeds
Mar	Jovaní	Universitat Politècnica de Catalunya
Sandeep	Kumar	Basque Center for Applied Mathematics
Daniel	López	Universitat Autònoma de Barcelona
Fernando	Martinez	Universitat Politècnica de Catalunya
Marija	Mihova	Ss. Cyril and Methodius University
Alejandro	Molero	Universitat Autònoma de Barcelona
James	Morrill	University of Oxford

M. Aguareles et al. (eds.), *Applications of Industrial Mathematics*,
RSME Springer Series 8, https://doi.org/10.1007/978-3-031-32130-6

Timothy	Myers	Centre de Recerca Matemàtica
Joaquim	Ortega	Universitat de Barcelona
Juan R	Pacha	Universitat Politècnica de Catalunya
Marta	Pellicer	Universitat de Girona
Mohammad	Peyman	Universitat Autònoma de Barcelona
Fouad eddine	Refisse	University of Djelfa
Jordi	Ripoll	Universitat de Girona
Antonio	Rodriguez	Universitat Politècnica de Catalunya
Rahil	Sachak-Patwa	University of Oxford
Daniel	Salgado	Universitat Autònoma de Barcelona
Jordi	Saludes	Universitat Politècnica de Catalunya
Aleix	March	Universitat Autònoma de Barcelona
Isabel	Serra	Centre de Recerca Matemàtica
Joan	Solà-Morales	Universitat Politècnica de Catalunya
Wilfredo	Sosa	Centre de Recerca Matemàtica
Silviya	Varbanova	University of Veliko Turnovo
Aleix	Ruiz de Villa	Universitat Autònoma de Barcelona
Ridvan	Yaprak	Karadeniz Technical University

The manufacturer's authorised representative in the EU is Springer Nature Customer Service Centre GmbH, Europaplatz 3, 69115 Heidelberg, Germany. If you have any concerns regarding our products, please contact ProductSafety@springernature.com

Printed and bound by CPI Group (UK) Ltd, Croydon, CR0 4YY
28/09/2026
02224085-0015